Validation and Qualification in Analytical Laboratories

Ludwig Huber

Interpharm Press, Inc.
Buffalo Grove, Illinois

Invitation to Authors

Interpharm Press publishes books focused upon applied technology and regulatory affairs impacting Healthcare Manufacturers worldwide. If you are considering writing or contributing to a book applicable to the pharmaceutical, biotechnology, medical device, diagnostic, cosmetic, or veterinary medicine manufacturing industries, please contact our Director of Publications.

Social Responsibility Programs

Reforestation

Interpharm Press is concerned about the impact of the worldwide loss of trees upon both the environment and the availability of new drug sources. Therefore, Interpharm supports global reforestation and commits to replant trees sufficient to replace those used to meet the paper needs to print its books.

Pharmakos-2000

Through its Pharmakos-2000 program, Interpharm Press fosters the teaching of pharmaceutical technology. Under this program, complimentary copies of selected Interpharm titles are regularly sent to every College and School of Pharmacy worldwide. It is hoped that these books will be useful references to faculty and students in advancing the practice of pharmaceutical technology.

Library of Congress Cataloging-in-Publication Data

Huber, Ludwig, 1948–
    Validation and qualification in analytical laboratories / Ludwig Huber.
       p.   cm.
    Includes bibliographical references and index.
    ISBN 1-57491-080-9
    1. Chemistry, Analytic—Quality control.   I. Title.
QD75.4.Q34H83   1998
543—dc21   98-36511
                                                                                                                                      CIP

10 9 8 7 6 5 4 3

ISBN: 1-57491-080-9
Copyright © 1999 by Interpharm Press, Inc. All rights reserved.

All rights reserved. This book is protected by copyright. No part of it may be reproduced, stored in a retrieval system, or transmitted in any form or by any means, electronic, mechanical, photocopying, recording, or otherwise, without written permission from the publisher. Printed in the United States of America.

    Where a product trademark, registration mark, or other protected mark is made in the text, ownership of the mark remains with the lawful owner of the mark. No claim, intentional or otherwise, is made by reference to any such marks in this book.

    While every effort has been made by Interpharm Press, Inc., to ensure the accuracy of the information contained in this book, this organization accepts no responsibility for errors or omissions.

Interpharm Press, Inc.
1358 Busch Parkway
Buffalo Grove, IL 60089, USA
Phone: + 1 + 847 + 459-8480
Fax: + 1 + 847 + 459-6644

Contents

Preface		ix
1.	**Introduction**	1
2.	**Regulations, Standards and Guidelines**	9
	Overview	10
	Specific Regulations and Guidelines	11
	Specific Quality Standards and Guidelines	18
	Guidance Documents of National and International Organizations	19
	Advice from Experts	23
	How to Deal with Multiple Regulations and Quality Standards	23
	Summary Recommendations	25
3.	**Terminology and Validation Strategy**	27
	Definitions	28
	Validation Versus Verification, Testing, Calibration and Qualification	31
	Strategy for Development and Implementation of a Qualification and Validation System in a Laboratory	35
	Summary Recommendations	39
4.	**Design Qualification**	41
	Recommended Steps in Design Qualification	42
	Case Study	43
	Vendor Qualification	43
	Summary Recommendations	47
5.	**Installation Qualification**	49
	Preinstallation	50
	Installation	51
	Tests During Installation	53
	The Installation Qualification Protocol	57

	Requalification After Changes to the System	59
	Summary Recommendations	59
6.	**Operational Qualification**	**61**
	Considerations	63
	Documentation	69
	A Practical and Economical Approach for Implementation	70
	Summary Recommendations	72
7.	**Performance Qualification and Maintenance**	**73**
	Logbook	75
	Maintenance	77
	Calibration	79
	Performance Testing	80
	System Suitability Testing	82
	Quality Control Samples with QC Charts	83
	Handling of Defective Instruments	89
	Summary Recommendations	91
8.	**Operational Qualification of Software and Computer Systems**	**93**
	Introduction	93
	Computerized Analyses Systems	95
	Computer Network	99
	Existing Systems and Systems Without Vendor Validation	100
	User-Contributed Software (e.g., Macros)	103
	Implementation and Documentation	104
	Summary Recommendations	105
9.	**Validation of Analytical Methods**	**107**
	Introduction	107
	Strategy for the Validation of Methods	110
	Validation of Standard Methods	115
	Validation of Nonroutine Methods	119
	Quality Control Plan	120
	Implementation to Routine Analysis	121
	Revalidation	122
	Parameters for Method Validation	123
	Summary Recommendations	140

10.	**Data Validation and Evaluation of Uncertainty**	**141**
	Validation of Data	*142*
	Reporting Data	*144*
	Measurement and Reporting of Uncertainty	*146*
	Summary Recommendations	*150*
11.	**(Certified) Reference Standards**	**151**
	Introduction	*151*
	Applications of (Certified) Standards	*153*
	Types of Material and Definitions	*153*
	Regulatory and Standard Requirements	*155*
	Official Reference Material Programs	*156*
	Traceability to National or Other Well-Characterized Standards	*158*
	Requirements for (Certified) Reference Material	*160*
	Preparation and Testing of (Certified) Reference Material	*161*
	Preparation of "Homemade" Reference Standards	*162*
	Correct Use of Certified Reference Material	*164*
	Quality Assurance Program	*165*
	Availability of (Certified) Reference Material	*165*
	Summary Recommendations	*168*
12.	**People**	**169**
	Recruiting Qualified People	*170*
	Defining and Communicating Job Descriptions, Tasks, Responsibilities and Desired Outcome	*178*
	Training	*179*
	Summary Recommendations	*188*
13.	**Proficiency Testing for External Laboratory Qualification**	**189**
	Procedure	*190*
	Evaluation of Proficiency Testing	*191*
	Who Should Participate in Proficiency Testing?	*192*
	Frequency of Tests	*192*
	Testing Material	*193*
	Advantages for Laboratories	*193*
	Performance Improvements	*194*

	Remaining Issues	194
	Summary Recommendations	195
14.	**Audits**	**197**
	Observations Reported During Inspections and Audits	198
	Planning and Implementation of Internal Audits	200
	Audit Report	203
	Audit Checklist	205
	Summary Recommendations	205
Appendix A. Glossary		**211**
Appendix B. OQ Tests for Selected Equipment		**229**
	Traceability of Standards	229
	Acceptance Limits	229
	Documentation and Archiving	229
	Gas Chromatography (GC)	230
	Capillary Electrophoresis	231
	UV/Visible Spectrophotometer	232
	High Performance Liquid chromatography	234
	Infrared/Near Infrared	237
	Analytical Balance	238
	Flame Atomic Absorption Spectrophotometer	238
	Laboratory Ovens	239
	Laboratory Furnaces	239
	Sterilizers (Hot Air)	240
	Refrigerators and Freezers	240
	Thermometers and Thermocouples	241
	Karl Fisher Apparatus	242
	Analog/digital Converter (from Ref 97)	243
	Dissolution Testing	244
	Viscosimeter	245
	Melting Point	246
	pH meter	246
	Refractometer	247
	Polarimeter	247
	Declaration of Operational Qualification	249

Appendix C. Selected (Standard) Operating Procedures 251
General Recommendations 251
Types and/or Content of SOPs 254
Proposal for a Title Page 256
General Workflow Diagram of SOPs for Equipment Testing 257
Example #1: SOP for the Preparation of Standard Operating Procedures 258
Example #2: Validation of Analytical Methods 261
Example #3: Testing Precision of Peak Retention Times and Areas of an HPLC System 270
Example #4: Retrospective Evaluation and Validation of Existing Computerized Analytical Systems 275

Appendix D. Selected Case Studies for OQ 283
Scenario 1: Pharmaceutical QC Lab with 25 Existing HPLC Systems 283
Scenario 2: Environmental Testing Lab Equipped with One HPLC System 285

Appendix E. Books in the Area of Qualification and Validation 287
Quality Assurance/Quality Control 287
Computer Validation 290
Method Validation/Statistics 292
ISO 9000 293
Food Laboratory 294
Regulations and Guidelines—GLP, GALP, cGMP, GCP, . . . 294
Pharmaceutical 297

Appendix F. References 299

Index 309

Preface

Validation and qualification for analytical methods and equipment are required by many regulations, quality standards and company policies. If executed correctly, they can also help to improve the reliability, consistency and accuracy of analytical data. *Validation and Qualification in Analytical Laboratories* guides analysts, laboratory managers and quality assurance managers through the validation and qualification processes in analytical laboratories.

> The main purpose of this book is to answer the key question regarding validation: How much validation is needed and how much is sufficient?

This book takes into account most national and international regulations and quality standards. Its concept, examples, templates and operating procedures are based on my multinational experience and incorporate all aspects of validation and qualification used at Hewlett-Packard and taken from personal discussions with regulatory agencies, managers and chemists in laboratories, corporate quality assurance managers and vendors of equipment and chemicals. Inputs have also arisen from discussions during Hewlett-Packard's worldwide seminar series covering all aspects of qualification and validation. Readers of this book will learn how to speed up their validation and qualification process, thereby avoiding troublesome reworking and gaining confidence for audits and inspections.

The validation and qualification procedures presented in this book help to ensure compliance and quality but with minimal extra cost and administrative complexity. Its purpose is to answer the key question regarding validation: *How much validation is needed and how much is sufficient?* The recommendations are complementary rather than contradictory to any standards or official guidelines. They are based mainly on common sense and can be used in cases where information from official guidelines and standards is insufficient for day-to-day work.

The concepts and ideas expressed in this book are my own and do not necessarily reflect official Hewlett-Packard policy.

Regulations and guidelines, and even more so their interpretations, are sometimes updated, and new ones are developed. Certain information in this book may, therefore, become incomplete with time. Reprint updates are limited, on a practical basis, to time periods of 3 to 5 years, a time span that suffices for over 90 percent of the content of this book. However, a timely update of all information is possible only using new, on-line information tools, such as the Internet. To take this fact into account, I have set up a special Web home page dedicated to this book:

http://www.t-online.de/home/huberl/validat.htm

You can find updated information under "book." As we have all experienced, Web addresses may change over time, sometimes with no influence from the owner. If this happens, I will try to maintain a link to the new address. If you are unable to find this link, please send me an e-mail, to 07802981947-0001@t-online.de or a fax or letter to one of the numbers or addresses given below.

On the Web page, you will also find many references to useful information relevant to this book but could not be included because of limited space. For example, it includes links to the pages of regulatory agencies that sometimes provide a free download of official documents.

In 1995, I wrote a book on another validation topic that also was published by Interpharm Press: *Validation of Computerized Analytical Systems*. By 1996, the book had become Interpharm's Number 1 bestseller, and I have received a lot of good feedback. Readers especially liked the bulleted lists, checklists, templates and Standard Operating Procedures. In this new book, I have incorporated and further elaborated on this concept. Those of you who possess a copy of my last book will find that the titles of some of the chapters overlap, for example, the chapter on the validation of methods. I received many requests to write a separate book on all aspects of validation, which I have done with this publication, and to update my previous book on computerized analytical systems, giving more specific information on software and computer validation. The figure below illustrates the concept of both books:

Interpharm

Validation of Computerized Analytical Systems

Ludwig Huber 1995

Reference 49

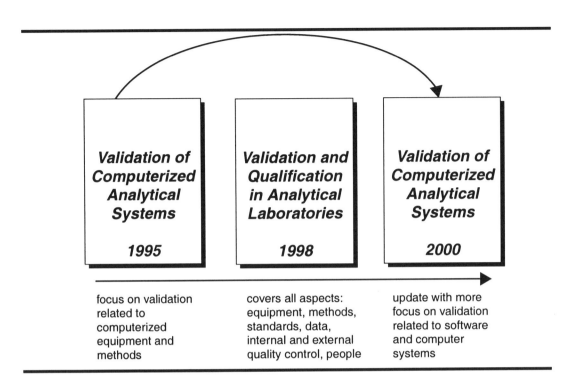

Comments

This book is intended to help clarify certain current issues in the area of validation and qualification in analytical laboratories. Readers are encouraged to submit their comments and suggestions, especially if their experiences have been different in daily laboratory work. Comments should be submitted to the author.

>Ludwig Huber
>Hewlett-Packard Waldbronn Analytical Division
>P.O. Box 1280
>76337 Waldbronn
>Germany
>Fax.: +49 7243 602501 or +49 7802 981948
>Phone.: + 49 7243 602209
>E-mail: ludwig_huber@hp.com or 07802981947-0001@t-online.de
>URL: www.t-online.de/home/huberl/validat.htm

Ludwig Huber
September 1998

1. Introduction

The purpose of any chemical analytical measurement is to get consistent, reliable and accurate data. There is no doubt that incorrect measurement results can lead to tremendous costs.

- If a product with incorrectly measured specifications is marketed, it may have to be recalled.

- If harmful contaminants in environmental samples are not detected, they can be dangerous to the environment.

- If drugs with undetected impurities are distributed, they can have a negative impact on peoples' health.

In addition, reporting incorrect analytical results at any particular time leads to loss of a laboratory's confidence in the validity of future results. Therefore, any laboratory should do its utmost to ensure measuring and reporting reliable and accurate data within a known level of confidence. Validation and qualification of processes and equipment will help meet this goal.

There is a second aspect to why validation and qualification are important, and this is equally important for those working in a regulated and in an accredited environment. Even though frequently not directly spelled out in regulations and official guidelines, such as Good Laboratory Practice (GLP) and Good Manufacturing Practice (GMP), or in accreditation standards, such as the International Organization for Standardization (ISO) Guide 25 or European Norm (EN) 45001, validation and qualification are usually required, which is confirmed by typical statements such as this one that appears in the U.S. cGMP (current Good Manufacturing Practice) regulations [1]: "Equipment shall be routinely calibrated, inspected and checked according to a written program to ensure proper performance." Failing a regulatory inspection can have an immense impact on a company; for example, in the pharmaceutical industry, marketing of new products may be delayed.

Because of their importance, validation issues have been addressed by several public and private organizations:

- The U.S. Food and Drug Administration (FDA) has published a technical guide on the *Validation of Chromatographic Methods* [2].

- The U.S. Environmental Protection Agency (EPA) has developed the *Guidance for Methods Development and Methods Validation for the Resource Conservation and Recovery Act (RCRA) Program* [3].

- The Association of Official Analytical Chemists (AOAC) has as one of its primary objectives the development and publication of analytical methods for substances affecting public health and safety, economic protection of the consumer or quality of environment and has published guidelines on method validation as part of *The Peer-Verified Methods Program* [4].

- ISO has developed guides on control charts [5–8] for data validation and on the qualification of reference material [9–11].

- The U.S. pharmaceutical industry has focused much attention on computer system validation, in particular in manufacturing environments. The U.S. Pharmaceutical Manufacturers Association (PMA) established the Computer System Validation Committee (CSVC) to develop guidelines for the validation of computer systems already in the early 1980s [12–13].

- The UK Pharmaceutical Analysis Science Group (PASG) has developed a position paper on the qualification of analytical equipment [14].

- The Laboratory of the Government Chemist (LGC) and EURACHEM-UK have developed a guidance document with definitions and step-by-step instructions for equipment qualification [15].

US FDA

Technical Review Guide

Validation of Chromatographic Methods, Center for Drug Evaluation and Research (CDER)

1993

Reference 2

EURACHEM-UK LGC-VAM

Guidance on Equipment Qualification of Analytical Instruments

December 1996

Reference 41

Validation is an old concept in analytical laboratories. Any good scientist has always validated an analytical method before using it for routine analysis, and equipment has been tested before it has been used for measurements. Therefore, the reader may ask, "Why is there a need for such a book at all?" In today's analytical laboratories, there are many problems with validation and qualification. Some of these problems are as follows:

1. Frequently, there is a lack of documented procedures and documented validation results. Analysts have often told this author that their methods and equipment have been validated, but when asked for reference plots or for the exact procedure they used, there was no documentation available. The subject of validation has brought about this major change: the need for documenting procedures and the validation results.

2. Only a part of the total analytical procedure has been validated but not the complete procedure. For example, frequently the analysis itself has been validated but not the sampling or sample preparation steps, which very often contributes most to an overall error (Figure 1.1).

3. Accessories and materials used for equipment qualification are not qualified. For example, there is no quality assurance (QA) program for chemical standards used to calibrate the equipment.

4. Procedures, performance parameters and acceptance limits for the operational qualification (OQ) of equipment hardware are not known.

5. There is a lot of uncertainty about procedures and the frequency of software and computer system validation.

6. Frequently, qualification and validation are done at just one particular point in time. A method is validated at the end of development, and equipment is qualified to meet specifications at the time of installation. However, validation and qualification are an ongoing process and cover the complete life of a method or equipment.

7. There is a lot of information and assistance from vendors on the qualification of newly purchased systems.

4   *Validation and Qualification in Analytical Laboratories*

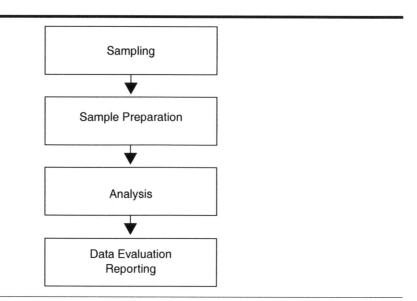

Figure 1.1. Validation activities should include the complete analytical procedure.

However, users of existing equipment are unsure if the same criteria should be used for the qualification of existing systems as for new systems (Figure 1.2).

This book is intended to help readers find answers to these problems. Recommendations made in the book reflect the author's common sense and are based on practical experience. They are not contradictory to any standards or official

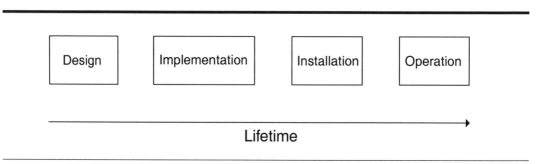

Figure 1.2. Equipment qualification covers the complete lifetime.

guidelines but complementary. They can be used in cases where information in official guidelines and standards is insufficient for day-to-day work. References to official guidelines and standards are given where appropriate.

This book covers all practical aspects of validation and qualification in analytical laboratories (Figure 1.3). Following a chapter on regulations, quality standards and related guidelines, it continues with terminology and an overview of the validation steps. The succeeding five chapters discuss the qualification of equipment, hardware and software, starting from design qualification (DQ) to installation qualification (IQ) to operational (OQ) and performance qualification (PQ).

The following three chapters are dedicated to the validation and qualification of methods, data and reference compounds. The final three chapters discuss the qualification of people, proficiency testing for external quality and audits.

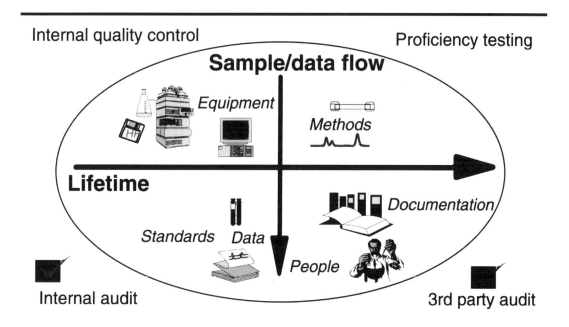

Figure 1.3. Validation and qualification in analytical laboratories cover all steps in the sample and data flow. They include procedures, methods, equipment throughout its entire lifetime, people and documentation.

The appendices include practical examples and procedures for validation and qualification, for example, a standard operating procedure for validation of analytical methods.

Even though the book uses mainly chromatography as an example, the concepts and strategies can be applied to the validation of other analytical techniques and equipment. The author has tried to cover as many aspects as possible and has made references to the relevant quality standards and regulations in the individual chapters. This does not mean that all recommendations should be followed for every analysis. The reader should carefully evaluate whether or not recommendations made in the book are appropriate for his or her work. Conclusions of the evaluation and their implementation in a laboratory should be part of an overall quality system and documented in a quality manual.

Regulations and quality standards as well as related guidelines are not specific enough and leave a lot of room for analysts, inspectors and auditors. If there is any doubt, the final answer can be obtained only by asking if the qualification or validation effort adds any scientific value. One should never forget that the primary goal of any analyst is to generate and deliver analysis data that are scientifically sound, whether they are submitted to a regulatory agency as part of a new drug application or delivered to a company's internal or external client. Well-designed, developed and validated analytical methods and equipment, together with motivated and qualified people, are prerequisites to achieve this goal and are part of Good Analytical Practice.

The challenge for any validation activity is to find the optimal validation effort that is somewhere between doing nothing and the attempt to validate everything to 100 percent. For example, the author has experienced these two extremes with the validation of commercial standard software for chromatographic instrument data evaluation. Some users felt they had to do nothing in the laboratory; others went to the full check of each function, which took several months, even when this was done already at the vendor's site.

The optimization process is illustrated in Figure 1.4. When done right at the beginning of the validation process, the additional value of each validation step is tremendous.

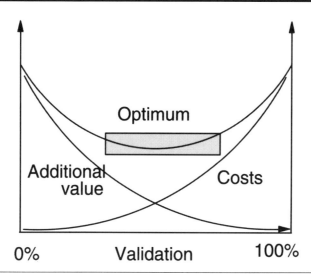

Figure 1.4. Optimization of validation, additional value and cost versus completeness of validation.

However, when trying to validate everything, the additional value goes to zero. On the other hand, the incremental costs for validation go up with any validation effort. The question is: "Where is the optimum?" or "How much validation is enough?" The challenge is to find the optimum. The hope is that, with the help of this book, the reader gets enough of a guide to finding this optimum for his or her specific process, which depends not only on the complexity of the tasks but also to some extent on the industry and country.

2. Regulations, Standards and Guidelines

What will be discussed in this chapter?

1. Regulations and guidelines that may be important for analytical laboratories (GLP, cGMP, ICH, USP)

2. Quality and accreditation standards that may be important for analytical laboratories (ISO Guide 25, EN 45001, United Kingdom Accreditation Service [UKAS])

3. The key contents of regulations and quality standards regarding validation and qualification

4. Guidance documents that are available for interpreting the regulations and guidelines

5. Books and other important publications that are available from experts

All regulations and quality standards that are applied in analytical laboratories include one or more sections on validation, verification or qualification of reference material, equipment, methods or procedures. Their general requirement is "suitability for intended use," which means, in practice, qualified or validated to meet previously specified requirements.

Legislation is one of two major forces driving validation and qualification in analytical laboratories. The second and more important reason for validation and qualification is to improve analytical results. It has always been an objective of good analytical scientists to meet this second goal, and, in this respect, validation and qualification are nothing new in analytical laboratories. In every case, a good scientist should have checked his or her equipment for performance level and should have validated analytical methods and procedures before using them routinely. Although most individuals have been qualified for their jobs, documented procedures have frequently not been followed for, and results have not been accurately documented. These two important points are invariably reviewed by regulatory inspectors and ISO 9000 auditors.

If validation and qualification are not performed according to regulations and guidelines, laboratories will fail to pass inspections or audits. Therefore, regulatory requirements and quality standards play a major role in all validation and qualification issues, and everyone working on regulated or quality standards should be familiar with their requirements. This entire chapter has been dedicated to regulations and standards because of their importance and impact on analytical laboratories. The most significant regulations and guidance documents are also described in an overview. Certain specific regulations are discussed in more detail in consecutive sections.

Overview

The most important regulations applying to validations are the (current) GMPs and the GLPs. The best known quality standards are the ISO 9000 series, which provide generic standards for development, manufacturing and service. The most frequently used quality and accreditation standards in chemical testing laboratories are ISO/IEC (International Electrotechnical Commission) Guide 25 and EN 45001.

When the first regulations and quality standards were released, there was little guidance for chemists in analytical laboratories on how to apply these regulations to analytical equipment and procedures. Regulations often include inexplicit sentences, such as "equipment should be of appropriate design and adequate capacity and shall be adequately tested and calibrated." Laboratory chemists have been uncertain about exactly what the words *appropriate* design and capacity or *adequate* testing and calibration mean in practice. It was usual for chemists to perform a particular task and for internal and external auditors to come in and instruct them to do further or different tests and to prepare additional documentation. This often occurred without being backed by any specific scientific reasons. Lab managers and chemists were asked to satisfy all the inspectors' requirements in order to be absolutely certain to pass the next audit. For example, Sharp [16] reported a case where he had asked a senior technician in a major British company if he would accede to any U.S. FDA request or suggestion, no matter how unreasonable or absurd it was. The technician replied, "We would not argue. We would just do it. The U.S. market is too big to lose."

Validation and qualification practices are more strongly driven by the level of enforcement across industries and countries than by any noticeable differences in regulations. It is the author's experience that the highest enforcement level in industry can be seen in pharmaceutical manufacturing, closely followed by pharmaceutical development. From a geographical point of view, the highest enforcement level is found in the United States. Outside the United States, this occurs mainly in Japan and in the United Kingdom, followed by other European countries and Canada.

This is an unsatisfactory situation and leaves a lot of room for uncertainty. One question that has frequently arisen is: "How much validation is enough?" The real problem lies in the lack of clear guidelines on how to implement regulations and quality standards. What kind of testing and how much testing is required are left very open to the interpretation of the internal and external auditors.

This situation is changing somewhat. Regulatory agencies have developed inspection guides and other guidance documents that are now available to the public. Even though these documents are not legally binding, they are used during inspections to check the compliance of the site to regulations.

Examples of such guidance documents in the area of cGLP and GMP are as follows:

- U.S. FDA Inspection Guides
- U.S. FDA Policy Guides
- Handbook for Inspectors in Germany

The U.S. FDA makes most of these guidance documents available to the public through the Internet (address: http://www.fda.gov/cder/guidance/index.htm).

Other agencies have also developed guidance documents for specific topics. For example, the U.S. EPA has developed the Good Automated Laboratory Practice (GALP) recommendations [17], with its focus on using computers in a regulated environment.

Private organizations, sometimes funded by the public, have developed guidance documents on selected topics. Examples are given below:

- The International Conference for Harmonization (ICH) has developed guidance documents for the validation of analytical procedures and other topics [18, 19].

- The U.S. Pharmacopeia (USP) has developed guidelines on validation of analytical methods and system suitability testing [20].

- The WELAC (Western European Laboratory Accreditation Conference)/EURACHEM has developed an interpretation guide on EN 45001 and ISO Guide 25 [21] and a guide on quantifying uncertainty in analytical measurement [22].

- The UK Pharmaceutical Analysis Science Group (PASG) has developed a position paper on equipment qualification for pharmaceutical laboratories [14].

- The UK Laboratory of the Government Chemist (LGC) and EURACHEM have developed guidance documents on equipment qualification [15].

- The Co-operation on International Traceability in Analytical Chemistry (CITAC) has developed an international guide to quality in analytical chemistry [23].

This chapter will discuss in more detail the content of these regulations and quality standards and related guidelines regarding validation and qualification in analytical laboratories.

Specific Regulations and Guidelines

Good Laboratory Practice

GLP regulations for assuring the validity of toxicological studies were first proposed by the U.S. FDA in November 1976, and final regulations were codified as Part 58 of

Chapter 21 of the *Code of Federal Regulations* in 1978 [24]. For safety testing of agricultural and industrial chemicals under the Federal Insecticide, Fungicide and Rodenticide Act (FIFRA) [25] and the Toxic Substance Control Act (TSCA) [26], respectively, the EPA issued almost identical regulations in 1983 to cover required health and safety aspects. The Organization for Economic Cooperation and Development (OECD) published the principles *Good Laboratory Practice in the Testing of Chemicals* [27] in 1982, which has since been updated [28] and incorporated by OECD member countries into their own legislation. In Europe, the European Community (EC) has made efforts to harmonize laws through council directives on *The Harmonization of Laws, Regulations and Administrative Provisions to the Application of the Principles of Good Laboratory Practice and the Verification of Their Application for Tests on Chemical Substances* (1987) [29] and on *The Inspection and Verification of Good Laboratory Practice* (1988, adopted in 1990) [30]. To overcome trade differences and enable GLPs to be recognized abroad, bilateral memoranda of understandings (MOUs) were developed.

All GLP regulations include chapters on equipment design and maintenance, for example, U.S. GLP regulations, Sections 58.61 and 58.63 [24]:

> Automatic, mechanical, or electronic equipment used in the generation, measurement, or assessment of data shall be of appropriate design and adequate capacity to function according to the protocol and shall be suitably located for operation, inspection, cleaning, and maintenance.
>
> Equipment used for generation, measurement, or assessment of data shall be adequately tested, calibrated, and/or standardized.
>
> Written standard operating procedures shall set forth in sufficient detail the methods, materials, and schedules to be used in routine inspection, cleaning, maintenance, testing, calibration, and/or standardization of equipment and shall specify remedial action to be taken in the event of failure or malfunction of equipment.
>
> Written records shall be maintained of all inspection operations.

The GLP principles of the OECD include similar but shorter sections on equipment [27]:

- The apparatus used for the generation of data and for controlling environmental factors relevant to the study should be suitably located and of appropriate design and adequate capacity.

- Apparatus and materials used in a study should be periodically inspected, cleaned, maintained, and calibrated according to Standard Operating Procedures. Records of procedures should be maintained.

Current Good Manufacturing Practice Regulations

> **US FDA**
> **21 CFR Part 211**
> Current Good Manufacturing Practice of Certain Requirements for Finished Pharmaceuticals
> **Proposed Rule May 1996**
>
> Reference 52

Good Manufacturing Practice (GMP) regulates manufacturing and its associated quality control (in contrast to GLP, which mainly covers drug development activities). GMP predates GLP. Industries were already familiar with GMP and thus GLP takes a similar line; the most significant difference is in archiving requirements for test samples and data.

GMP regulations have been developed to ensure that medicinal (pharmaceutical) products are consistently produced and controlled according to the quality standards appropriate to their intended use. In the United States, the regulations are called current Good Manufacturing Practices (cGMP) to account for the fact that the regulations are dynamic rather than static. They are defined in Title 21 of the U.S. Code of Federal Regulations, 21 CFR 210—*Current Good Manufacturing Practice for Drugs, General*—and 21 CFR 211—*Current Good Manufacturing Practice for Finished Pharmaceuticals*. Drugs marketed in the United States must first receive FDA approval and must be manufactured in accordance with the U.S. cGMP regulations. Because of this, FDA regulations have set an international regulation benchmark for pharmaceutical manufacturing.

In Europe, local GMP regulations exist in many countries. These are based on the EU directive: *Good Manufacturing Practice for Medicinal Products in the European Community*. This EU GMP is necessary to permit free trade in medicinal products between the member countries. Regulations in the EU allow the marketing of a new drug in the member countries with the acquisition of just a single marketing approval. The intention of the EU GMP is to establish a minimum manufacturing standard for all member countries.

The EU directive has been widely harmonized with the *Guide to Good Manufacturing Practice for Pharmaceutical Products* as developed by the Pharmaceutical Inspection Convention (PIC).

The English text of 27 national and international (current) GMPs can be found in the book *International Biotechnology, Bulk Chemical, and Pharmaceutical GMPs* (5th ed.) [31]. International GMPs include versions from the World Health Organization (WHO), Asia, the PIC and the EU.

GMP is concerned with both production and quality control (QC). The basic requirements of QC are as follows:

- Adequate facilities, trained personnel and approved procedures are available for sampling, inspecting and testing starting materials, packaging materials, intermediate bulk and finished products and, where appropriate, for monitoring environmental conditions for GMP purposes.

- Samples of starting materials, packaging materials, intermediate products, bulk products and finished products are taken by personnel and by methods approved by QC.

- Test methods are validated.

- Records are made manually and/or by recording instruments that demonstrate that all required sampling, inspecting and testing procedures were actually carried out. Any deviations are fully recorded and investigated.

- The finished products contain active ingredients complying with the qualitative and quantitative composition of the marketing authorization, are of the purity required and are enclosed within their proper container and correctly labeled.

- Records are made of the results of inspection and demonstrate that testing of materials and intermediate, bulk and finished products is formally assessed against specification. Product assessment includes a review and production documentation relevant to evaluation and an assessment of deviations from specified procedures.

- No batch of a product is released for sale or supply prior to certification by an authorized person to confirm that it is in accordance with the requirements of the marketing.

- Sufficient reference samples of starting materials products are retained to permit future examination of the product if necessary so that the product can be retained in its final pack, unless exceptionally large packs are produced.

- Sufficient referee samples of starting materials products are retained to permit future examination of the product, if necessary; the product can be retained in its final pack, unless exceptionally large packs are produced.

The U.S. FDA published a *Guide to Inspection of Pharmaceutical Quality Control Laboratories* [32]. Even though it was written as a guideline for field investigators, it is, nevertheless, a useful document for QC laboratories. It includes extensive chapters on the handling of "Failure (out of specification) of Laboratory Results" and on "Retesting." Additional chapters provide guidelines on laboratory records and documentation, laboratory standards solutions, methods validation, equipment, raw material testing, in-process control, the computerized laboratory data acquisition system and laboratory management.

The *EC Guide to Good Manufacturing Practice* [33] contains one short section on equipment and method validation:

1. All equipment should be subject to planned maintenance and validation.

2. Analytical methods should be validated. All testing operations described in the marketing authorization should be carried out according to the approved methods.

Appendix 11 of the EC guide has more specific information on the use of computers in the GMP environment and includes sections on validation.

International Conference on Harmonization Guidelines

The International Conference on Harmonization of Technical Requirements for Registration of Pharmaceuticals for Human Use is a unique project that brings together the regulatory authorities of Europe, Japan and the United States, and the experts from these three regions, to discuss scientific and technical aspects of product registration [34]. The ICH has three purposes:

1. To provide a forum for a constructive dialog between regulatory authorities and the pharmaceutical industry on the real and perceived differences in the technical requirements for product registration in the EU, United States and Japan. Members come from industry and regulatory agencies.

2. To identify areas where modifications in technical requirements, or greater mutual acceptance of research and development procedures, could lead to a more economical use of human, animal and material resources without compromising safety.

3. To make recommendations on practical ways to achieve greater harmonization in the interpretation and applications of technical requirements for registration.

> **ICH**
> Validation of analytical procedures: definitions and terminology
> 1996
>
> Reference 18

The ICH publishes the results as guidelines to regulatory authorities and to industry in the member countries. The member countries are finally supposed to sign off the guidelines to produce regulations. The most important guidelines related to the topic of this book are stability testing and the validation of analytical methods. Two guidelines have been published on method validation to date:

1. Q2A: Validation of analytical procedures [18], with a list of performance criteria for method validation and definition of the terminology

2. Q2B: Validation of analytical procedures: methodology [19] with recommendations on how to measure and evaluate some of the parameters

U.S. Pharmacopeia

The USP is the official compendium recognized by the U.S. Federal Food, Drug, and Cosmetic Act. It serves as the basis for enforcement actions by the U.S. FDA, involving official (USP) drugs, and also provides guidelines during foreign inspections. It contains chapters on the validation of analytical methods and system suitability testing [20, 35].

Specific Quality Standards and Guidelines

Most chemical analytical laboratories already possess, or are in the process of implementing, a quality management system to improve the quality, consistency and reliability of data. A documented quality system is also a prerequisite for obtaining accreditation or for registering to a quality standard such as ISO 9001, 9002 or 9003.

ISO 9000 Series Quality Standards and ISO 9000-3

> **ISO 9000-3**
>
> Guidelines for the application of ISO 9001 to the development, supply and maintenance of software
>
> 1991
>
> Reference 38

The quality standards ISO 9001 through ISO 9003 cover the requirements for a generic quality system in a two-party contradictory situation with an assessment made by a third party. The standards are neither specific to laboratory work nor to computer systems or software. Recognition that the development, manufacturing and maintenance processes of software are different from those for most other products led to the issuance in 1991 of ISO 9000-3: *Guidelines for the Application of ISO 9001 to the Development, Supply and Maintenance of Software* [38]. This provides guidance for quality systems involving software products and deals with situations where specific software is developed, supplied and maintained according to a purchaser's specification as part of a contract.

ISO/IEC Guide 25 and EN 45000 Series

> **ISO/IEC Guide 25**
>
> General requirements for the calibration and competence of testing laboratories
>
> 1990
>
> Reference 37

EN 45001:1989 (*General Criteria for the Operation of Testing Laboratories*) [36] and ISO/IEC Guide 25 (*General Requirements for the Competence of Calibration and Testing Laboratories*) [37] are frequently used as guides in establishing a quality system in chemical testing laboratories. Both documents may be used as a basis for laboratory accreditation.

EN 45001 and ISO/IEC Guide 25 both include a chapter on equipment, similar to the equipment sections found in the GLP and cGMP regulations.

TickIT and ITQS

Software development and maintenance activities can be formally assessed using the TickIT and Information Technology Quality System (ITQS) [39] scheme through the application of ISO 9000-3. The TickIT scheme was investigated by the Department of Trade and Industry in the United Kingdom. The

TickIT guide [40] comprises 172 pages with the following chapters:

- Introduction
- ISO 9000-3: *Guidelines for the Application of ISO 9001 to the Development, Supply and Maintenance of Software*
- Purchaser's Guide
- Supplier's Guide
- Auditor's Guide (includes the European IT Quality System Auditor Guide)

The guide complements ISO 9000-3 by providing additional guidance on implementing and auditing a quality management system (QMS) for software development and support.

ITQS [39] also checks compliance with ISO 9000-3, and its scope is similar to TickIT. A major objective of ITQS is to expand the international agreement on the registration of IT companies and to provide a global scheme for the recognition of certification in this area. It has been developed by the British Standards Institute and other European Standard Organizations.

Both TickIt and ITQS registration are extensions of ISO 9001 in the area of software and computer systems and may be required by some companies as part of the software vendor qualification program.

>
> **ITQS**
>
> **European Information Technology Quality System Auditor Guide**
>
> **1992**
>
> Reference 39

Guidance Documents of National and International Organizations

Most national standard, accreditation or certification bodies issue guidance notes in support of their standards. For example, in the United Kingdom, the National Measurement and Accreditation System (NAMAS) has published more than 100 such documents, many of which relate to chemical testing. Complete lists and ordering details are generally available on request from various bodies.

Co-operation on International Traceability in Analytical Chemistry

CITAC devised the *International Guide to Quality in Analytical Chemistry—An Aid to Accreditation* [23] with the intent of

> **CITAC**
>
> **International Guide to Quality in Analytical Chemistry— An Aid to Accreditation**
>
> **1995**
>
> Reference 23

providing laboratories with guidance on the best practice for improving the quality of the analytical operations they perform. The document was developed from the EURACHEM/ WELAC Guide [21] and has been updated to account for new material and developments and views from outside Europe and to reflect an approach less closely associated with accreditation or certification. The guide was produced by a CITAC working group with representatives from the testing industry, EURACHEM, EAL (European Co-operation for the Accreditation of Laboratories), ILAC (International Laboratory Accreditation Conference), AOAC, IUPAC (International Union of Pure and Applied Chemistry) and NIST (U.S. National Institute of Standards and Technology). It is comprehensive and very detailed concerning all aspects of validation and qualification of equipment.

- All equipment used in laboratories should be of specification sufficient for the intended purpose and kept in a state of maintenance and calibration consistent with its use.

- Equipment normally found in the chemical laboratory can be categorized as:

    - general service equipment not used for making measurements or with minimal influence (e.g., hotplates, stirrers, nonvolumetric glassware and glassware used for rough volume measurements such as measuring cylinders) and laboratory heating or ventilation systems;

    - volumetric equipment (e.g., flasks, pipettes, pycnometers, burettes, etc.) and measuring instruments (e.g., hygrometers, U-tube viscosimeters, thermometers, timers, spectrometers, chromatographs, electrochemical meters, balances, etc.);

    - physical measurement standards (weights, reference thermometers); and

    - computers and data processors.

While these sections in the document are rather generic and similar to those found in other documents, the real value is contained in the appendices. For example, Appendix A includes a quality audit checklist with areas of particular importance in a chemical laboratory. There are checklist items on equipment, methods and QC:

Equipment

- The equipment in use is suited to its purpose.

- Major instruments are correctly maintained and records of this maintenance are kept.

- Appropriate instructions for the use of equipment are available.

- Traceable equipment (e.g., balances, thermometers, glassware, timepieces, pipettes, etc.) are appropriately calibrated, and the corresponding certificates or other records demonstrating traceability to national measurement standards are available.

- Calibrated equipment is appropriately labeled or otherwise identified to ensure that it is not confused with uncalibrated equipment and to ensure that its calibration status is clear to the user.

- Instrument calibration procedures and performance checks are documented and available to users.

- Instrument performance checks and calibration procedures are carried out at appropriate intervals and show that calibration is maintained and day-to-day performance is acceptable. Appropriate corrective action is taken where necessary.

- Records of calibration performance checks and corrective action are maintained.

Methods and Procedures

- In-house methods are fully documented, appropriately validated and authorized for use.

- Alterations to methods are appropriately authorized.

- Copies of published and official methods are available.

- The most up-to-date version of the method is available to the analyst.

- Analyses (are observed to) follow the methods specified.

- Methods have an appropriate level of advice on calibration and quality control.

Quality Control

- There is an appropriate level of quality control for each test.

- Where control charts are used, performance has been maintained within acceptable criteria.

- QC check samples are being tested by the defined procedures at the required frequency, and there is an up-to-date record of the results and actions taken where results have exceeded action limits.

- Results from the random reanalysis of samples show an acceptable measure of agreement with the original analyses.

- Where appropriate, performance in proficiency testing schemes and/or in interlaboratory comparisons is satisfactory and has not highlighted any problems or potential problems. Where performance has been unsatisfactory, corrective action has been taken.

Appendix B gives guidance on the performance checks and calibration intervals of equipment most commonly used in analytical laboratories. These include balances, hydrometers, barometers, timers, thermometers, gas chromatographs, liquid chromatographs and spectrometers.

Laboratory of the Government Chemist/ EURACHEM-UK

> **EURACHEM-UK**
> **LGC-VAM**
>
> *Guidance on Equipment Qualification of Analytical Instruments*
>
> **December 1996**
>
> Reference 41

The LGC established a working group, under the auspices of EURACHEM-UK, that developed a detailed guidance on equipment qualification. The group defined the individual qualification terms and gave recommendations on what should be included in each qualification. The guidance has been published with comments by Bedson and Sargent [15]. The same group is currently developing even more specific guidelines on the qualification of specific instruments, such as gas chromatographs and liquid chromatographs [41].

Analytical Instrument Association

The Analytical Instrument Association (AIA) has developed a voluntary guideline [42] for IQ (installation qualification), OQ

> **UK Pharmaceutical Analytical Sciences Group (PASG)**
>
> *Qualification of Analytical Equipment Position Paper*
>
> 1995

(operational qualification) and PQ (performance qualification) to assist their members in helping customers to comply with regulatory requirements concerning IQ, OQ and PQ.

The UK Pharmaceutical Analytical Sciences Group

The PASG developed a position paper on equipment qualification, along with broad guidelines as to what each qualification step should include [14]. The group introduced the terms *design qualification, installation qualification, operational qualification* and *performance qualification* for equipment in analytical laboratories.

Advice from Experts

Because of the lack of detailed information from regulatory agencies and other official organizations, individual experts or expert groups have emerged and have published books and other literature giving guidelines on validation and qualification. Appendix E of this book includes a bibliography of more than 30 books with titles, keywords and ordering information.

How to Deal with Multiple Regulations and Quality Standards

Laboratories are frequently faced with a situation where they have to comply with regulations from different countries or with both regulations and quality standards at the same time (Figure 2.1). Examples are as follows:

1. A pharmaceutical company markets a drug in different countries. Manufacturing and manufacturing control has to comply with the cGMP of all countries. In this case, the analytical control laboratory also has to work in compliance with the GMPs of the countries in which the drug is marketed.

2. A chemical company is certified for ISO 9001. The scope of the certification also covers the analytical service laboratory. In addition, the laboratory performs contract analyses for other companies and has received laboratory

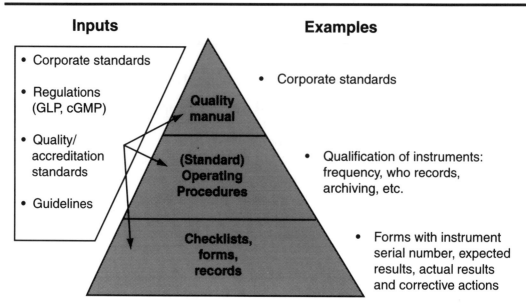

Figure 2.1. Quality pyramid of a quality system for multiple regulations and quality standards.

accreditation in compliance with ISO Guide 25. The laboratory has to work in compliance with ISO 9001 and with ISO Guide 25.

3. An independent test laboratory performs GLP studies as a subcontractor for a pharmaceutical company. Occasionally, the laboratory also performs analyses for pharmaceutical manufacturing control departments. The laboratory has also received laboratory accreditation for specific food analyses according to ISO Guide 25. The laboratory has to comply with ISO Guide 25 and with GLP and cGMP regulations.

International companies frequently face this kind of problem. Their laboratories not only have to comply with regulations from different countries but also, simultaneously, with quality and accreditation standards. The solution to this problem is to combine all regulations and quality standards in a single quality manual and a single set of operating procedures. The quality manual should place the company's own quality system first and foremost. This may be based on a well-known quality standard, such as ISO 9001. The quality manual and operating procedures should include aspects of various regulations and quality standards applied within the company. For specific regulations, such as GLPs, it should include sections that apply only to those particular regulations. For example, it might mention that if the analysis is to be done for a GLP study, raw data must be archived for the required archiving period. For a non-GLP type of analysis, such long archiving is not usually required.

Summary Recommendations

1. Check to determine which regulations and quality standards apply to your laboratory.

2. Try to procure inspection, policy and interpretation guides that are relevant to your laboratory.

3. Develop a quality manual for your entity that covers all regulations, quality standards and guidelines that are relevant to your laboratory.

4. Develop operating procedures and work instructions.

5. Inform yourself on current audit and inspection practices in your country.

3. Terminology and Validation Strategy

What will be discussed in this chapter?

1. The difference between validation, verification, qualification and calibration
2. Other definitions related to validation
3. The equipment qualification process
4. The elements required for a complete qualification and validation
5. How to develop and implement a validation strategy for your laboratory

An agreement on terminology is of utmost importance for a common understanding of validation and qualification. The author has frequently noted at validation symposia that different speakers used different terms for the same thing and the same terms for different things. Consequently, discussions would start on the topic of terminology that not only wasted valuable symposium time but also left some uncertainty, because official definitions were usually not readily available for clarification, and the speakers could not reach a consensus.

The main problem is that guidelines on validation and qualification have been developed by different organizations, at different times and in different countries. For example, the pharmaceutical industry uses the term *equipment operational qualification* while ILAC uses the term *equipment verification* for confirming an instrument's compliance with previously defined specifications. Frequently, the terms *validation* and *verification* or *validation* and *qualification* are used interchangeably.

This chapter elaborates on terms most frequently used in the area of validation and qualification (Figure 3.1) in analytical laboratories. Whenever available, official terms are used together with a reference to the source.

Definitions

The term *validation* has been defined in the literature by many different authors. Although the wording is different, the sense is always the same: (a) specify and implement, (b) test if the specifications are met and (c) document.

One of today's commonly accepted definitions of validation can be found in the guideline *General Principles of Validation* from 1987 [43]:

> Establishing documented evidence which provides a high degree of assurance that a specific process will consistently produce a product meeting its predetermined specifications and quality attributes.

This definition is very well thought-out, and each word has a special significance. Most important in this definition are the words *documented, high degree of assurance, specific process, consistently* and *predetermined specifications*.

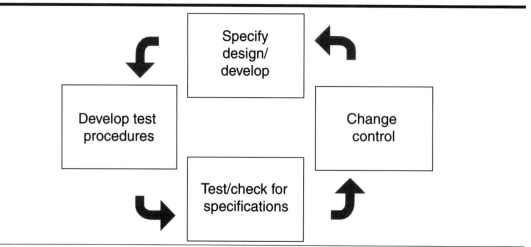

Figure 3.1. Principle of qualification and validation.

documented evidence	Validation requires thorough documentation. Everything that is not documented is considered incomplete.
high degree of assurance	The assumption is that a large software package as used in complex computerized systems, for example, is rarely free of errors. Frequently, there is a perception that validation means "error free." This assumption is wrong. During the validation process, everything realistically possible should be done to reduce errors to a "high degree."
specific process	The overall validation of software is process, not product, related. For example, the development and testing activities before releasing the software for manufacture are validated once for a series of products characterized by the serial number. Some subparts of validation, such as the qualifications (installation, operational, performance), are product specific and have to be done for each system.
consistently	Validation is not a one-time event. The performance of the equipment has to be controlled during the entire life of the product.
predetermined specifications	Validation activities start with the definition of specifications. The performance of the equipment is then verified against these specifications. Acceptance criteria must be defined prior to testing.

The GMP guides of the EC [33], the WHO [44] and the PIC [46] have the following definition for *validation:*

> Action of proving, in accordance with the principles of Good Manufacturing Practice, that any procedure, process, equipment, material, activity or system actually leads to the expected results.

CITAC's *International Guide to Quality in Analytical Chemistry* [23] and the EURACHEM/WELAC's *Accreditation for Chemical Laboratories: Guidance on Interpretation of the EN 45000 Series of Standards* and *ISO/IEC Guide 25* [21] define *validation of data and equipment* in Appendix C1.11 as follows:

> The checking of data for correctness, or compliance with applicable (of data processing) standards, rules and conventions. In the context of equipment rather than data, validation involves checking for correct performance, etc.

The OECD consensus document number 10 [47] defines *validation of a computerized system* as "the demonstration that a computerized system is suitable for its intended purpose."

The USP [36] defines *validation of analytical methods* as "the process by which it is established, by laboratory studies, that the performance characteristics of the method meet the requirements for the intended applications."

Many laboratory managers associate validation with increased workload in the laboratory or increased paperwork, but validation is essentially nothing new. Ever since the development of analytical instrumentation and methods, statistics have been used to prove the proper functioning, reliability and precision of the equipment and methods. Firms followed software development standards and used the software development life cycle long before regulatory agencies requested the validation of computer systems. New to most existing validation procedures is the disciplined planning of validation and documentation of all validation steps, including testing. This also concurs with a definition of validation by Ken Chapman [48]:

> In today's pharmaceutical industry, whether you are thinking about a computer system, a water treatment system, or a manufacturing process, validation means nothing else than well-organized, well-documented common sense.

Validation Versus Verification, Testing, Calibration and Qualification

There is still considerable misunderstanding on the differences between testing, calibration, verification and validation. The illustration in Figure 3.2, together with other information in the following paragraphs, should help to clarify these differences.

Testing

Testing has been defined in ISO/IEC Guide 25 as

> A technical operation that consists of the determination of one or more characteristics or performance of a given product, material, equipment, organism, physical phenomena, process or service according to a specified procedure.

Instrument testing is the process of executing experiments to measure the performance characteristics following documented procedures. Examples are the measurement of the baseline noise of a detector, the precision of the injection

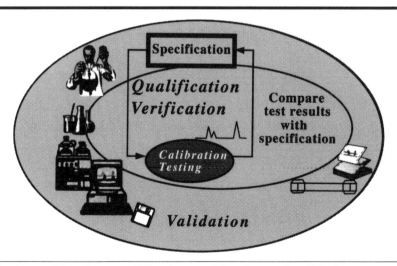

Figure 3.2. Testing, calibration, qualification, verification and validation.

volume of an injector or the precision of a flow rate. Requirements for testing are test conditions and Standard Operating Procedures (SOPs) with clear instructions on how to do the tests and how to evaluate the results.

Calibration

ISO/IEC Guide 25 defines *calibration* as

> The set of operations which establish, under specified conditions, the relationship between values indicated by a measuring instrument or measuring system and the corresponding known values of the measurand.

A well-known example of a device that has to be calibrated is the balance. A reference weight that is traceable to a national standard is measured and the result compared with the actual weight. An example for a calibration procedure in an analytical instrument is the measurement and adjustment of the wavelength accuracy in a high performance liquid chromatography (HPLC) UV-visible detector's optical unit. Calibration is frequently confused with testing and performance verification. The differences become quite clear when looking at the precision of the peak area: This can be tested and verified against a previously defined specification, but it cannot be calibrated. Sometimes, accurate calibration has a direct impact on performance. For example, a UV detector with incorrect wavelength calibration may cause detection limits and the detector's linearity to deteriorate. The term *calibration* is sometimes used interchangeably with the term *standardization*. Calibration normally means to check against known standards, whereas *standardization* usually means to make uniform [69]. For some equipment, the term *calibrated* is more appropriate; for other equipment, the term *standardized* is better [69].

Verification

ISO/IEC Guide 25 defines *verification* as the "confirmation by examination and provision of evidence that specified requirements have been met." Performance verification of analytical instrumentation is the process of comparing the test results with the specification. It includes testing and requires the availability of clear specifications and acceptance criteria. Examples are the same as for testing. The verification process ends with the generation and sign-off of a "Declaration of Conformity" of the instrument to specifications. Additionally, a sticker should be affixed to the instrument with the date of the last successful performance verification and the next scheduled performance verification.

The international guidelines for the development, supply and maintenance of software (ISO 9000-3) [38] differentiate between verification and validation. *Verification* is defined as

> The process of evaluating the products of a given phase to ensure correctness and consistency with respect to products and standards provided as input to that phase.

Validation is defined as "the process of evaluating software to ensure compliance with specified requirements." Validation relates to the complete product, whereas verification is related to the individual phases or modules.

Qualification

The term *qualification* has been defined by the U.S. PMA's CSVC for the installation, operation and running of a system under workload for a specific application. Like verification, qualification is also part of validation and is product specific. The CSVC has defined three qualifications: installation, operational and performance.

Figure 3.3 illustrates the qualification timeline. It demonstrates that validation is not a one-time event but an ongoing process starting with the definition and design of the product.

Qualification also is defined, in the *EC Guide to Good Manufacturing Practice* [33], as the "action of proving that any equipment works correctly and leads to the expected results." The word *validation* is sometimes widened to incorporate the concept of qualification.

Even though the term *qualification* has been used in analytical laboratories, it was formally introduced by a workgroup of the UK PASG in a poster paper [14] at the PharmAnalysis Europe conference in 1995, which was published in 1995 by PharmTech Europe. The authors applied the terms for qualification (IQ, OQ and PQ), which were previously applied to

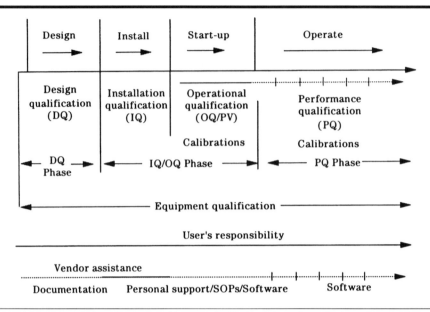

Figure 3.3. The qualification timeline, originally developed for the validation of a computer-controlled water treatment system, is now applied to analytical equipment used in analytical laboratories.

the qualification of computer systems by the U.S. PMA, to analytical equipment. The authors also introduced the term *design qualification* (DQ).

The UK LGC defined the individual qualification terms and gave detailed recommendations on what should be included in each qualification. The guidelines, together with some comments, have been published by Bedson and Sargent [15].

Equipment qualification (Figure 3.4) has been broken down into four parts:

1. DQ, for setting functional and performance specifications (operational specifications);

2. IQ, for performing and documenting the installation in the selected user environment;

3. OQ, for testing the equipment in the selected user environment to ensure that it meets the previously defined functional and performance specifications; and

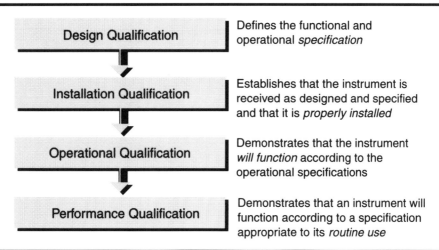

Figure 3.4. Equipment qualification.

4. PQ, for testing that the system performs as intended for the selected application.

The standardization of terminology amongst suppliers and users of analytical equipment is necessary. In response to this need, the AIA developed a *Voluntary Guideline for IQ, OQ and PQ* [42] to assist their members in helping customers comply with regulatory requirements concerning IQ, OQ and PQ. They used mainly the terminology as developed by the LGC/EURACHEM working group [15]. Unfortunately, *design qualification* was not defined by the AIA.

Table 3.1 summarizes situations where different equipment qualification steps are required.

Strategy for Development and Implementation of a Qualification and Validation System in a Laboratory

Validation efforts in an analytical laboratory can be broken down into separate components addressing the validation/qualification of

- equipment hardware,
- software and computer systems,

Table 3.1. Needs for equipment qualification

	DQ	IQ	OQ	PQ
Before purchasing	yes			
During installation		yes		
Before operation			yes	yes
During operation			(2)	yes
After hardware repair, e.g., replace GC injection port		(1)	partial	yes
Hardware update, e.g., additional LC detector	partial	partial	partial	yes
Firmware update		partial		yes
Software update	(3)	partial	(3)	yes
Move equipment to other building		partial	yes	yes
New operator				yes
Column replacement (in chromatography)				yes
New use of equipment (new application not previously specified)	yes		(4)	yes

(1) Only if the part to be exchanged has new serial number.

(2) Frequency depends on the equipment, for example, for a chromatograph, it is about once per year.

(3) Yes, if the update includes new functions that will be used for the user's application.

(4) Yes, if new functions are used that previously have not been tested.

- analytical procedures and methods,
- analytical system,
- analytical data,
- reference standards and
- people.

The various validation activities in an analytical laboratory are illustrated in Figure 3.5.

For overall validation and qualification as well as validation processes, the author recommends the following steps:

1. Develop procedures for validation and qualification.
2. Make sure that all laboratory work staff is adequately qualified through appropriate education, training or experience. Training needs should be established, and records of the staff's qualifications should be kept.

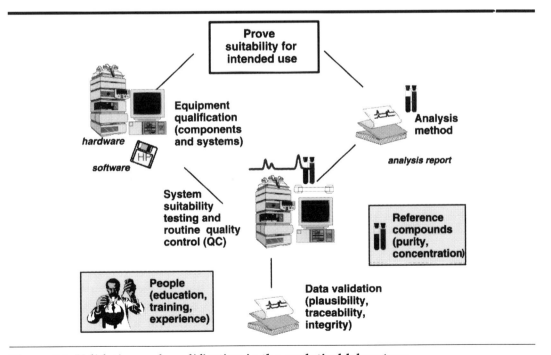

Figure 3.5. Validation and qualification in the analytical laboratory.

3. Qualify tools and chemical standards for instrument calibration and QC checks following documented plans. Such qualification can include verification of the traceability of calibration tools to national standards and the amounts and purity of chemical standards. Define the intended use and specifications for equipment and analytical procedures.

4. Qualify analytical hardware at installation prior to routine use and, if necessary, after repair and at regular intervals.

5. Validate software and computer systems during and at the end of the development process. If such systems are purchased, the vendor should be qualified.

6. Qualify software and computer systems at the time of installation in the user's laboratory prior to their routine use, at regular intervals and, if necessary, after software and hardware updates.

7. Validate analytical methods during and after development. Method validation covers definition and testing of significant method characteristics, for example, limit of detection, limit of quantitation, selectivity, linearity and ruggedness. If the method is to be run on different instruments, it also should be validated on the different instruments as specified in the scope of the method. Only when it is clearly specified that the method will always run on the same instrument can validation efforts be limited to that instrument. Methods should be validated at the end of method development prior to routine use and whenever any method parameter has been changed.

8. Combine a specific method with specific equipment hardware, software, accessories (such as columns) and chemical standards, and test the suitability of the system for a specific analysis. This qualification, usually referred to as PQ, tests a system against documented system performance specifications for the specific analytical method. Analytical systems should be tested for ongoing performance prior to and during routine use, practically on a day-to-day basis.

Terminology and Validation Strategy 39

9. When analyzing samples, validate the data. The validation process includes documentation and checks for data plausibility, data integrity and traceability. The uncertainty of the measurement results should be estimated and reported together with the analytical data. A complete audit trail that allows the final result to be traced back to the raw data should be in place.

10. Verify that the entire analytical procedure is validated with well-characterized control samples that are interspersed between unknown samples. Results of control samples are compared with known amounts. If the results are within specified limits, the complete analytical procedure is validated. This process is called internal analytical QC.

11. For external analytical QC, participate in proficiency testing. Well-characterized samples with known amounts are distributed to a group of laboratories doing similar analyses. The samples are analyzed, and the results are sent back to the distributing agency. The agency evaluates the results and informs the laboratories of their performance.

12. Conduct regular internal audits to check that the laboratory's QA system is effective, documented and adhered to by the entire staff.

The details of these steps will be discussed in consecutive chapters in this book.

Summary Recommendations

1. Develop a glossary with terms on regulations for your entity. Appendix A of this book includes an extensive glossary on all aspects of validation and qualification in analytical laboratories. This glossary can be used to develop a company-specific glossary.

2. Make sure that the definitions that you have developed for your entity are used throughout all processes, in all documents and in all departments.

3. Make sure that all employees have a copy of the definitions that you have developed for your entity readily

available when they attend external meetings and conferences.

4. Develop and implement a validation and qualification strategy for your laboratory.

4. Design Qualification

What will be discussed in this chapter?

1. Definition of design qualification (DQ)
2. Importance of DQ
3. Steps for DQ
4. How vendors can help with DQ
5. Steps for vendor qualification

> Design qualification defines the functional and operational specifications of the instrument and details the conscious decisions in the selection of the supplier [15].

DQ should ensure that instruments have all the necessary functions and performance criteria that will enable them to be successfully implemented for the intended application and to meet business requirements. Errors in DQ can have a tremendous technical and business impact, and, therefore, a sufficient amount of time and resources should be invested in the DQ phase. For example, setting wrong operational specifications can substantially increase the workload for OQ testing, and selecting a vendor with insufficient support capability can decrease instrument up-time with a negative business impact.

While IQ, OQ and PQ are being performed in most regulated laboratories, DQ is a relatively new concept to many laboratories. It is rarely performed in those cases where the equipment is planned for use in multiple applications, not in a specific one. When the author has presented the concept of equipment qualification in seminars, attendees always agreed that the concepts of IQ, OQ and PQ are essential for analytical laboratories, but many have not been convinced that this also holds true for DQ. However, the author is totally convinced that equipment cannot be qualified without DQ. DQ is a prerequisite of IQ and OQ, and most users today give "must have" specifications when they purchase equipment from a vendor, which is part of DQ. The main purpose of DQ is to ensure that

- the right type of equipment is selected for specific tasks,
- the equipment will have the right functional and performance specifications and
- the vendor meets the user firm's qualification criteria.

DQ should be performed

- when a new instrument is being purchased or
- when an existing instrument is being used for a new application not previously specified.

Recommended Steps in Design Qualification

Table 4.1 lists recommended steps that should be considered for inclusion in a design qualification.

Table 4.1. Steps in design qualification

- Describe the intended use of the analysis.
- Select the analysis technique.
- Describe the intended use of the equipment.
- Describe the intended environment.
- Define user requirement specifications (URS).
- Preliminarily select the functional and performance specifications (technical, environmental, safety).
- Preliminarily select the supplier.
- Test the selected instrument (if the technique is new).
- Finally, select the equipment and supplier.
- Document the final functional and operational specifications.

It is frequently the case that instruments are used for different applications with different functional and performance requirements. In this case, the recommendation is to describe the most important intended applications and to specify the functional and performance specifications so that they meet the criteria for all applications. It is also possible to develop a generic DQ for instrument categories that will be used for similar applications.

To set functional and performance specifications, the vendor's specification sheets can be used as guidelines. However, the author would not recommend simply writing up the vendor's specifications because compliance to the functional and performance specifications must be verified later on in the process during OQ and PQ. Specifying too many functions and setting the values too stringently will significantly increase the workload for OQ.

Case Study

Table 4.2 includes a case study for pesticide analysis in drinking water. The parameters can be easily adapted to other applications. The table can be used as a basis for writing any design qualification.

Vendor Qualification

As part of the DQ process, the vendor should be qualified: The question is, how should this be done? Is an established and documented quality system enough (e.g., ISO 9001)? Should there be a direct audit? Is there another alternative, between these two extremes?

It is the author's opinion that this item is frequently and heatedly discussed by consultants. There may be situations where a vendor audit is recommended, for example, when complex computer systems are being developed for a specific user. However, this is rarely the case for analytical equipment. Typically, off-the-shelf systems are purchased from a vendor with little or no customization for specific users.

In most cases, it is also a question of confidence between the vendor and the user's firm. Unfortunately, confidence is not taken into account in arguments with the FDA. So how

Table 4.2. Selected examples for a design qualification of an HPLC system (incomplete)

Design Qualification	Selected Examples
Intended use	Analysis of pesticides in drinking water with quantitation limit of 10 ppt/compound.
Sample preparation and analysis technique	Sample cleanup and enrichment with solid phase extraction. HPLC for analysis.
User requirement for HPLC analysis	• 20 samples/day. • Automated unattended sample injection, HPLC analysis, data evaluation and printing. • Limit of quantitation: 1 pg/peak. • Automated confirmation of peak identity and purity with diode-array detection.
Functional specifications	
Pump	• Binary or higher gradient. • Flow rate: 0.5 to 5 mL/min. • Must have on-line vacuum degasser.
Detector	• UV/VIS diode-array. • Must monitor 3 signals at the same time. • Wavelength range: 200 to 600 nm. • Flow cell and lamp must be accessible from the front for easy maintenance. • Detector must have holmium oxide filter for automated wavelength calibration.
Autosampler	• Must accommodate at least 100 samples with 2 mL volume or at least 25 samples with 5 mL volume. • Variable sample volume from 1 µL to 5 mL without hardware change. • Needle flush and wash to minimize sample carryover.

Continued on next page.

Continued from previous page.

Design Qualification	Selected Examples
Column compartment	• Operating range 20 to 40°C, peltier controlled.
	• Must accommodate at least 2 columns with length of up to 25 cm.
System	• Ability to detect leaks in each module and to switch the pump off in case a leak is detected.
	• Maintenance parts accessible from the front.
Computer and software	• Microsoft NT operating system.
	• Control of all module and system parameters.
	• Data acquisition for signals and spectra, peak integration and quantitation, spectral evaluation for peak purity and compound confirmation.
	• Automated method sequencing for unattended injection of different samples.
	• Electronically save and retrieve all chromatograms and method parameters generated and used by the system.
	• Database for storage of and search for data.
	• Selection and analysis of control samples and database with on-line plot of selected QC parameters. System must stop if control parameters are out of specified limits.
	• Recording of unusual events in an electronic logbook.
	• Software for routine diagnostics and troubleshooting hints.
Operational	• Detector: baseline noise: $< 5 \times 10^{-5}$ AU; wavelength accuracy: $< 2\%$.
	• Sampler inj. volume precision: $< 0.5\%$ RSD at 10 µL inj. volume; carry over: $< 0.5\%$.
	• Pump: precision of retent. time: $< 0.5\%$ RSD.
	• Column compartment: temp. precision $< 0.5°C$.

Continued on next page.

Continued from previous page.

Design Qualification	Selected Examples
User instructions	• Operating manual on paper. • Computer-based tutorial. • OQ procedures in SOP format.
Validation/qualification	Vendor must provide IQ and OQ services.
Site preparation	Vendor must provide instructions on environmental conditions, required space and any supplies required for the equipment.
Environmental conditions	Instrument must operate reliably at ambient temperatures from 20 to 35°C.
Maintenance	• Vendor must deliver maintenance procedure and recommend schedule. • Instrument must include early maintenance feedback for timely exchange of most important maintenance parts. • Vendor must provide list of maintenance parts with part numbers and prices. • Maintenance procedures must be supplied on multimedia CD-ROM.
Safety	System must include leak detection and drain for safe leak handling. Pump must stop delivering mobile phase if a leak is detected.
Training	Vendor must provide familiarization of the equipment hardware and software and training on how to maintain the equipment.
Service and support	Vendor must provide service contract.

should one proceed? The exact procedure depends very much on the individual situation.

- Does the system being considered employ mature or new technology?
- Is the specific system in widespread use either within your own laboratory or your company, or are there references in the same industry?
- Does the system include complex computer hardware and software?

For example, if the equipment does not include a computer system, certification to ISO 9001, or an equivalent system, is sufficient. When the assumption is that the equipment to be purchased is not an off-the-shelf commercial system that includes a computer for instrument control and data handling, the steps described in Table 4.3 are recommended, insofar as there is no previous experience with this vendor in your company.

Summary Recommendations

1. Take DQ more seriously.
2. Develop a procedure for DQ.
3. Develop a procedure for vendor qualification.
4. Define the intended use of the equipment.
5. Define the required functions of the equipment (use the vendor's instrument specifications list for help).
6. Qualify the vendor, based on references (internal and external) and/or by using mail audit checklists.

Table 4.3. Steps for vendor qualification (for systems not off-the-shelf)

1. *Develop a vendor qualification checklist.* This list should include questions on how the equipment is developed, validated, installed and supported (a more complete example of such a list is shown in Reference 49). The most important questions are as follows:

   - Does the vendor have a documented and certified quality system, e.g., ISO 9001 (please note: ISO 9002 or 9003 is insufficient because they don't cover development!)?
   - For products that include software: Does the vendor comply with ISO 9000-3 (*Guidelines for the Application of ISO 9001 to the Development, Supply and Maintenance of Software*) or an equivalent standard or guide?
   - Are equipment hardware and computer software developed and validated according to a documented procedure, e.g., according to a product life cycle?
   - Is the vendor prepared to make product development and validation records and source codes accessible to regulatory agencies?
   - Does the vendor provide declarations of conformity to documented specifications?
   - Does the vendor provide assistance in DQ, equipment installation, OQ, maintenance and timely repair through qualified people?
   - Is there a customer feedback and response system in case the user reports a problem or enhancement request?
   - Is there a change control system with appropriate notification of users subsequent to changes?

2. *Send the checklist to the vendor.* If the vendor answers all of the questions satisfactorily within a given time frame, the vendor is qualified.

3. *If the vendor does not answer the questions satisfactorily, another vendor should be considered.* If there is no other vendor who could provide an instrument that meets the operational and functional specifications, a direct audit should be considered.

5. Installation Qualification

What will be discussed in this chapter?

1. Definition of installation qualification (IQ)
2. Steps for IQ
3. The line between IQ and OQ
4. Documentation that should be generated
5. Performance of IQ—the vendor or the user

   Installation qualification establishes that the instrument is delivered as designed and specified, that it is properly installed in the selected environment, and that this environment is suitable for the operation and use of the instrument [15].

The main purposes of IQ are to ensure that the

- equipment has been received as purchased,
- the selected environment meets the vendor's environmental specifications,
- individual hardware modules and all accessories are properly installed and connected to each other,
- the software is completely installed on the designated storage device,
- the instrument functions in the selected environment and
- all equipment hardware and software are registered in some kind of a laboratory equipment database.

This chapter discusses issues that involve preparation for installation: installation of hardware and software, functional testing of modules and systems and preparing the documentation. IQ should follow a process that can be documented as a (standard) operating procedure.

Preinstallation

Before the instrument arrives at the user's laboratory, serious thought must be given to its location and space requirements. A comprehensive understanding of the requirements for the new equipment must be obtained from the vendor well in advance: required bench or floor space, environmental conditions such as humidity and temperature and, in certain cases, the utility needs such as electricity or compressed gases for gas chromatographs. Care should be taken that all of the environmental conditions and electrical grounding are within the limits specified by the vendor and that the correct cables are used. If environmental conditions might influence the validity of test results, the laboratory should have facilities to monitor and record these conditions when using the equipment, either continuously or at regular intervals. Examples are measurements of environmental temperature and humidity if the instruments are operated close to the specified limits.

Any special safety precautions should be considered (for example, for radioactivity measurement devices), and the location should also be checked for any devices generating electromagnetic fields in close proximity. Table 5.1 lists the recommended steps before installation.

Table 5.1. Steps before installation

- Obtain manufacturer's recommendations for installation site requirements.
- Check the site for the fulfillment of the manufacturer's recommendations (utilities such as electricity, water and gases; environmental conditions such as humidity, temperature, vibration level and dust).
- Allow sufficient shelf space for the equipment, SOPs, operating manuals and software.

Installation

When the instrument arrives, the shipment should be checked for completeness by the user. It should be confirmed that the equipment ordered is what was in fact received. In addition to the equipment hardware, other items should be checked, e.g., cables, other accessories and documentation. A visual inspection of the entire hardware system should follow to identify any physical damage. For more complex instrumentation, for example, if a single computer controls and/or acquires data from several analytical instruments, wiring diagrams should be produced if they were not supplied by the vendor. An electrical test of all modules and systems should follow. The impact of electrical devices close to the computer system should be considered and evaluated if the need arises. For example, when variable voltages are sent between sensors and integrators or computers, electromagnetic energy emitted by poorly shielded fluorescent lamps in close proximity or by motors can interfere with the transmitted data. Table 5.2 lists steps as recommended during installation.

When the installation procedure is completed, both hardware and software should be well documented with model, serial and revision numbers.

For larger laboratories with large amounts of equipment, a computer database for the storage of instrument records is preferable. Items that should be included for each piece of equipment are listed in Table 5.3.

Detailed documentation is even more important for computer systems than for equipment hardware. Documentation should include items such as the size of the hard disk, internal memory (RAM), installed type and version of operating software, standard application software and user contributed software (e.g., MACRO programs; see Table 5.4). This information is important because each item can influence the overall performance of a computer system. The information should be readily available when a problem occurs with the computer system.

Table 5.2. Steps during installation

- Compare equipment, as received, with purchase order, including
  — software,
  — accessories,
  — spare parts and
  — consumables.
- Check documentation for completeness:
  — operating manuals,
  — maintenance instructions,
  — standard operating procedures for testing and safety,
  — validation certificates and
  — health and safety instructions.
- Check equipment for any damage.
- Read the supplier's instructions for installation.
- Read the supplier's safety instructions, if any.
- Install hardware (computer, equipment, fittings and tubings for fluid connections, columns in HPLC and GC, power cables, data flow and instrument control cables) following the manufacturer's recommendation.
- Switch on the instruments and ensure that all modules power up and perform an electronic self-test. Check if the instrument does what is described in the supplier's documents. Record any deviations.
- Install software on computer following the manufacturer's recommendation.
- Verify correct software installation. For example, are all files loaded? Utilities for verifying the installation should be included in the software itself.
- Make a backup copy of software.
- Configure peripherals, e.g., printers and equipment modules.
- Identify and make a list with a description of all hardware; include drawings where appropriate.
- Make a list with a description of all software installed on the computer.
- List equipment manuals and SOPs.
- Prepare an installation report.

Table 5.3. Equipment hardware characterization

- Internal identification number (asset number)
- Description of the piece of equipment
- The manufacturer's name, address and phone number for service calls and service contract number, if there is one
- Serial number and firmware revision number of equipment
- Date received
- Date placed in service
- Current location
- Size, weight
- Condition when received, e.g., new, used, reconditioned
- List with authorized users and responsible person(s)

Tests During Installation

One question that frequently arises is whether any type of testing should be done as part of IQ. Frequent questions include: Should there be any test? Should testing include functional, operational and performance measurements? Operational testing does not belong in IQ; it belongs in OQ. IQ should include tests only to verify that the software and hardware are installed properly and that all electrical and fluid connections are correct. Therefore, IQ should include switching on the instrument and checking for any error messages. Correct loading of computer software should be checked by suitable verification software. For a system that consists of several modules, such as a modular HPLC system, IQ can include injection and qualitative evaluation of a standard. In this way, the correct installation of all fluid and electrical tubings and cables can be checked.

Hardware Modules

Modern equipment hardware modules typically include a self-test program. They are performed every time the instrument is switched on; Table 5.5 lists typical electronic self-tests

Table 5.4. Form for computer system identification

Computer Hardware	
Manufacturer, model	
Serial number	
Processor	
Internal memory (RAM)	
Graphics adapter	
Hard disk (partitions, memory sizes)	
Installed drives	
Pointing device (e.g., mouse)	
Space requirement	
Monitor	
Manufacturer, model	
Serial number	
Printer	
Manufacturer, model	
Serial number	
Space requirement	
Instrument Interface Card	
Type, select code, slot number	
Connected Equipment Hardware	
Hardware module 1	
Interface card setting	
Network	
Type	
Operating Software	
Operating system (version)	
User interface (version)	

Continued on next page.

Continued from previous page.

Application Software 1	
Description	
Manufacturer/vendor	
Product number (version)	
Required disk space	
Application Software 2	
Description	
Manufacturer/vendor	
Product number (version)	
Required disk space	

Table 5.5. Typical electronic self-tests during instrument startup

- *Read Only Memory (ROM) test:* This test is performed automatically every time the instrument is started. It checks the integrity of the ROM processor by comparing the actual checksum number with the original checksum number burned into the ROM.

- *Random Access Memory (RAM) test:* Run during instrument startup, a series of numbers are written to and read from the processor RAM memory. Both series of numbers must be identical to pass this test.

- *Display test:* To ensure that all important user information is visible, the operation of all display devices, including LEDs, status and error lamps is checked.

- *Remote connections:* This tests the communications to and from external devices and checks their status: ready, not ready, error. This is an important function that enables any module to shut down the pumping device should a leak be detected anywhere in the system.

during instrument startup. The display of the message "self-test passed" is enough proof for successful installation (Figure 5.1).

Computer Systems

When complex software is installed on a computer, the correctness and completeness of the installed program and data files should be verified. Vendors can assist in this process by supplying installation reference files and automated validated verification procedures. In this case, the integrity of each file is verified by comparing the cross-redundancy check (CRC) of the installed file with the checksum of the original file recorded on the installation master. Modified or corrupt files have different checksums and are thus detected by the verification program. Verification reports should include a list of missing, changed and identical files (Figure 5.2).

Computerized Analytical Systems

Two critical installation items for a multimodule computerized system are the correct installation of fluid and electrical connections between different modules and the electrical connections between the computer hardware and the equipment hardware. This can be checked most efficiently by running a well-characterized sample through the system to acquire and

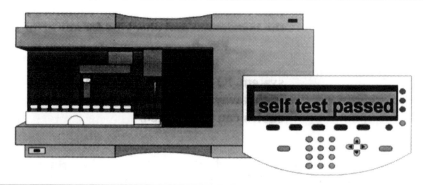

Figure 5.1. Equipment modules have a built-in electronic self-test program for instrument startup.

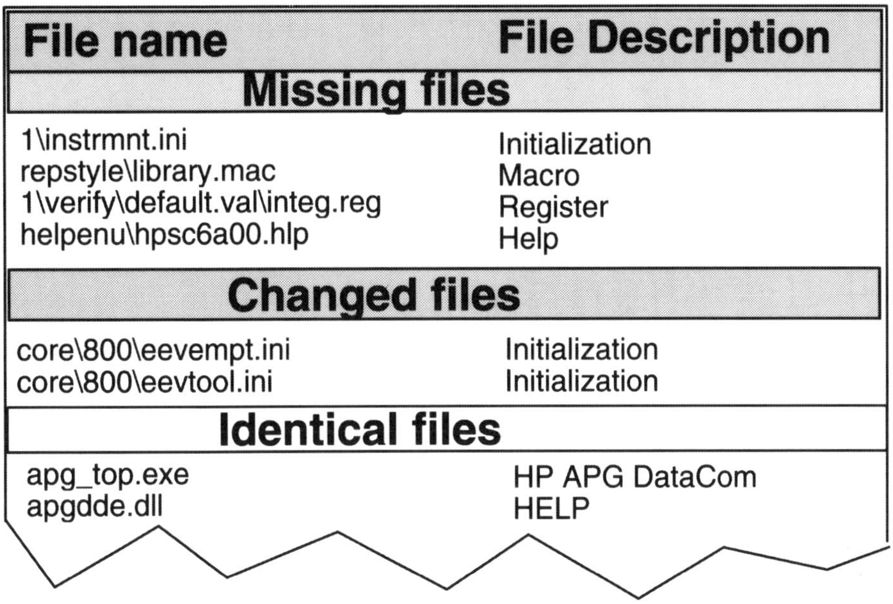

File name	File Description
Missing files	
1\instrmnt.ini	Initialization
repstyle\library.mac	Macro
1\verify\default.val\integ.reg	Register
helpenu\hpsc6a00.hlp	Help
Changed files	
core\800\eevempt.ini	Initialization
core\800\eevtool.ini	Initialization
Identical files	
apg_top.exe	HP APG DataCom
apgdde.dll	HELP

Figure 5.2. Software installation verification report.

evaluate the data using standardized methods and compare the computer printout (spectra or chromatograms) with reference plots. When the actual plots comply (agree) with the reference plots, this is enough evidence of a successful system installation (Figure 5.3).

The Installation Qualification Protocol

The installation should end with the generation and sign-off of the installation report referred to as the IQ document. The document should be signed by the user's representative, if the IQ was done by the user, and by the vendor's and the user's representative, if the IQ was done by the vendor.

It is recommended that documented procedures with pre-printed forms for the installation report be used. It is also recommended to make copies of all important documentation;

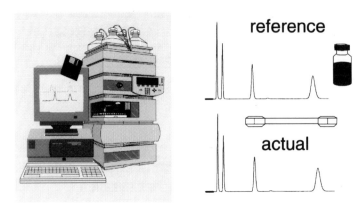

Figure 5.3. A system installation can be verified by analyzing a reference sample under well-characterized conditions and by comparing the actual plot with a reference plot.

one copy should be placed close to the instrument, and the other kept in a safe place. An identification sticker should be put on the instrument with information to include the instrument's serial number and the company's asset number.

The IQ protocol should include the following:

- Scope of the IQ protocol
- Protocol acceptance/approval by the user
- Protocol acceptance/approval by the vendor (if installation was performed by the vendor)
- Document revision history
- Listing of all instrumentation, including manufacturer, model and serial number, and so on
- List of user manuals and other documentation
- Procedure and results of module and/or system installation checks

Appendices should be attached that include the following:

- Purchase order
- Manufacturer's recommendation regarding the installation (they should include environmental limits, e.g., temperature and humidity), power and gas supply requirements
- Reference plots and printouts of results from installation testing
- Wiring diagram (for complex systems)
- Software status bulletin describing any known defects of the system and giving temporary work-around solutions

Requalification After Changes to the System

Any changes to the system should be thoroughly recorded and documented in order to maintain the qualification process. These may be replacements of modules or hardware or software upgrades. Depending on the change, the tests as described in the previous sections should be repeated, and the results should be documented.

Summary Recommendations

1. Develop an operating procedure for IQ.
2. Generate a database for equipment.
3. Ask the vendor to perform IQ as part of the installation.
4. Correct installation of software should be verified for computer systems. Develop an installation verification master file.
5. An installation check with known chemical standards should be performed for complex modular systems.
6. Document IQ. If IQ was done by the vendor, the IQ document should also be signed by the vendor.

6. Operational Qualification

What will be discussed in this chapter?

1. The steps required for operational qualification (OQ)
2. Selecting tests and acceptance criteria for OQ
3. How to document OQ
4. The criteria for requalification

   Operational qualification is the process of demonstrating that an instrument will function according to its operational specification in the selected environment [15].

For the qualification of computer systems, the terms *acceptance testing* and *performance verification* are also used. The main purpose of OQ is to ensure that

- the equipment's hardware meets functional and performance specifications as required for the intended application and as specified in the DQ document and

- the computer software meets functional specifications as required for the intended application and as specified in the DQ document.

OQ should be carried out

- after initial installation;
- after instrument repair and after other major events, such as upgrades; and
- at regular intervals during routine use.

OQ is an important part of the overall equipment qualification process. The careful selection of test items, the test procedures and acceptance limits is extremely important, because if set too stringently, the instrument's test may have an unnecessarily high failure rate and/or the maintenance efforts will be too intense. If the limits are too relaxed, the equipment will not prove itself fit for its purpose.

The general procedure to qualify an instrument for operation is as follows:

1. Define the intended use and the functional and operational specifications (use criteria as defined during DQ).
2. Develop test procedures.
3. Define acceptance criteria.
4. Perform the tests.
5. Document the results.
6. Develop criteria and steps for requalification, e.g., after repair.
7. Develop procedures in case the equipment does not perform to specifications.

The most difficult steps in this procedure are Step 3, some details of Step 4 and Step 6. While most analytical scientists today agree on the definition of OQ, agree on the general procedure for performing OQ and have some idea about the test procedure, they are still unsure about implementation. Questions also arise regarding requalification after instrument upgrade and repair or when the instrument is moved to another lab. Frequent questions regarding OQ are as follows:

1. What procedures and test standards should be used? Should they reflect the intended use of the equipment, or should they be generic for the instrument category?
2. What should the acceptance criteria be? Should they be in line with the manufacturer's specifications, or should they reflect the intended use of the equipment?
3. Should the same procedures and acceptance criteria be used for all instruments of the same type in my laboratory and/or in our company?
4. For modular systems, should each module be tested, or is it enough to test the system as a whole?

5. How should the computer and software part of a system be tested?

6. How frequently should the OQ tests be done?

7. Should the tests be redone after instrument upgrade, after a repair or when the instrument is moved to another lab?

8. Can or should the test be done by the vendor or by the user?

9. Can or should preventive maintenance be performed before the OQ test?

10. Why is OQ needed? If the equipment is used for one specific application only, isn't PQ enough?

In this chapter, the author gives recommendations related to the OQ of equipment hardware. OQ of software and computer systems will be discussed in Chapter 8. Case studies and test procedures for different types of equipment are available in Appendix D of this book.

Considerations

What Should Be the Test Items, Procedures and Acceptance Limits?

Before starting a discussion related to test items, test procedures and acceptance criteria, we should have a closer look at the OQ definition created by the LGC/EURACHEM working group:

> The process of demonstrating that an instrument will function according to its operational specification in the selected environment.

Another similar definition for OQ came from the U.S. PMA [50]:

> Documented verification that the equipment related system or subsystem performs as intended throughout representative or anticipated operating ranges.

Although this definition is brief and leaves a lot of room for interpretation, one thing becomes obvious: OQ should prove that the instrument is suitable for its intended use. OQ is not required to prove that the instrument meets the manufacturer's performance specifications. This is a frequent misunderstanding, yet many operators prefer to use the manufacturer's specifications because usually these are readily available.

However, a mistake such as this can have an enormous impact on the equipment's maintenance costs. One example is the baseline noise of a UV/visible detector, a performance criterion that is important for a method's limit of detection and limit of quantitation.

The baseline noise as offered today by many UV/visible detectors is in the range of 1 to 2×10^{-5} absorption units (AU) and much lower than the limits of detection and quantitation required for most applications. This value is achieved under optimum conditions, such as with a reasonably new lamp, an ultraclean flow cell, stable ambient temperature, HPLC grade mobile phase, no micro leaks in the entire HPLC system and so on. These conditions are always valid at the manufacturer's final test and probably at the time of installation in the user's laboratory. However, after some time, optical and mechanical parts deteriorate (e.g., the lamp loses intensity and the flow cell may become contaminated). So, if we repeat the test after 3, 6, or 12 months, the noise of 1×10^{-5} AU may no longer be obtained.

The question now is: How do we know when the detector was not within the OQ specifications? An auditor also may ask the question: How do we know that all the data measured in the past are valid if the instrument was not within the specifications as set by the user? In this case, it is therefore necessary to perform the OQ tests much more frequently, and to change the lamp more frequently, and probably clean the flow cell on a regular basis. This requires additional operator time and creates additional costs, which can be justified if the application requires low baseline noise. They cannot be justified if the instrument is used only for applications that don't require low baseline noise.

Which Test Sample Should Be Used: A Generic Standard or a Standard That Is Specific to the Application?

Let's assume the instrument is used for different applications, which require different samples, different columns and different calibration standards. In this case, it is recommended to use a generic standard for the same instrument category. It is also recommended to use the same approach if multiple instruments in a lab perform different applications. If there are just one or two instruments that run one type of application with one calibration standard, it makes sense to also use that standard for OQ.

Should All Instruments of the Same Category Meet the Same Criteria or Should Each Instrument Have Its Own Limits?

This is another question that comes up frequently in discussions. For example, you have in your lab HPLC systems from different vendors that may also have been purchased at different times. In this case, the instruments will have different performance characteristics. For example, the UV/visible detector's baseline noise has decreased by about a factor of 10 over the last 10 years. There may be instruments in the lab with 1×10^{-4} AU and others with 1×10^{-5} AU. The recommendation is to define 2×10^{-4} AU as a general limit. If there are applications on specific instruments that require a lower baseline noise, select the newer ones to be run for this application, and make an exception for the noise limit for this instrument, 2×10^{-5} AU, for example.

How Often Should OQ Be Performed?

Test frequency is another important question. Should it occur once a month, after several months or once a year? The answer depends on

- the type of equipment,
- the usage of the equipment,
- the nature of usage (environment, application),
- the stability of the equipment and
- the operational specifications set by the user.

The most important criterion here is to make sure that the test frequency selected will result in a high probability that the equipment will pass the tests.

Modular Versus Holistic Testing

Another frequent point for discussion used to be whether each individual module in a system should be tested (modular testing) or if the system should be tested as a whole (holistic testing) (Figure 6.1). This discussion was suddenly halted by a statement published by a U.S. FDA field investigator in 1994 [51]. The recommendation was, and still is, to generally use the holistic approach; test the system as a whole, and look into individual modules only if the system does not pass. However, this approach is more applicable for diagnostic purposes than for test purposes. The same recommendation has been made by the LGC/EURACHEM working group [15].

Who Should Perform the Test—a Representative of the Vendor or the User?

Test performance involves both resources and economics, in addition to the technical aspects of the test. In principle, testing can be done by both the user and the vendor. The technical question relates to the procedure the vendor offers: Does

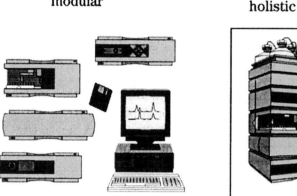

Figure 6.1. Modular versus holistic testing.

Standard Operating Procedures OQ/PV

Hewlett-Packard HP1100 Series HPLC

it really check the critical performance limits of the instrument? As long as test procedures relate to the intended use of the instrument, it may be more economical if a vendor does them. The advantage for the user is that he or she does not have to be careful about the traceability of tools such as thermometers, because the vendor's representative supplies everything. Also, for whatever reason, some auditors prefer to see a calibration stamp on the equipment that comes from outside the user's lab.

Should Preventive Maintenance Be Done Before OQ?

In our example with the UV/visible detector lamp, preventive maintenance would solve the problem of not being within specifications due to the lamp aging. The LGC/EURACHEM working group recommends performing preventive maintenance before OQ, yet some inspectors do not care for this procedure because there is no evidence that the instrument was performing properly at all times. Before a decision is made, one should think about the purpose of an OQ: Is it proof that the equipment did and does perform according to specification all the time, or should it make sure that the equipment is fit just for future tasks? The answer to this question will also answer the preventive maintenance question. In light of this, the exact purpose of OQ should be included in the operating procedure.

Maintenance & repair

Should the Tests Be Redone After Maintenance, Upgrade or Repair (Requalification)?

Whenever something has occurred with the equipment, be it a repair or an upgrade, the correct functioning and performance of the system should be verified through appropriate tests. This procedure is widely referred to as *requalification*. The type of tests depends on the type of repair or upgrade. For example, if an autosampler of an HPLC system has been repaired, the performance characteristics influenced by the autosampler should be tested—the injection volume precision and the carryover. There is no need to perform the pump's gradient accuracy or the detector's linearity. Instrument vendors should provide a list with recommendations for each system and module on what should be tested after a repair or a system upgrade. The tests used for requalification should be designed so that the results can be compared with those obtained from the initial qualification. Any significant differences in the results obtained from old and new tests should be identified, recorded and resolved.

What Should Be Done If a Module Is Replaced on a Modular System?

Sometimes complete modules are replaced either because a new application may require a new detector or because the old one is defective. In this case, all detector specific system tests, such as baseline noise, detector linearity or the wavelength accuracy of an HPLC UV detector, should be performed. It is not necessary to check other module parameters, such as the step gradient accuracy of the pump.

Should the Tests Be Redone When the Instrument Is Moved?

The retesting of equipment after it has been moved depends on the type of equipment and extent of movement. If the instrument is moved along a bench, no requalification is required for most instruments. For equipment with mechanical susceptibility to vibrations, part of the requalification is required. For example, for an HPLC system with a UV variable wavelength detector with a motor-driven grating, the wavelength accuracy should be verified. Similarly, if a balance is moved, it should be recalibrated. The situation is similar if the equipment is moved within a laboratory to another bench. If the instrument is moved to another laboratory within the same building or to another building with different environmental conditions, a full requalification should be performed.

What Should Be Done When the Instrument Is Used for Another Application?

If the application is similar and all OQ tests and acceptance criteria are covered through the initial OQ, no additional tests are required. If the new application has different or additional demands, which are not covered through the initial application, appropriate OQ tests should be done. For example, if for a chromatographic analysis the instrument initially was used for high concentrations with no need for baseline noise tests, such tests should be done if the new application requires the analysis of trace compounds. Other performance tests, such as the precision of the injection volume, are not required in this case.

Why Should I Do OQ at All? Isn't PQ Enough?

The final question that arises is: Why should I do OQ at all on a regular basis; isn't PQ enough? This is a valid question for many users. PQ has several advantages: It is done on a more frequent basis, and it is more specific to the user's application. If the instrument is used for just one or maybe only a few specific applications, and if the PQ tests include all relevant performance criteria, the regular OQ test may be omitted. The critical issue here is which parameters are tested within the PQ test.

One should also not forget that regular OQ tests provide ongoing information on the performance of the HPLC system. Performance trends can be measured and recorded and can also give an early indication that an instrument may no longer perform as expected in the near future. For example, gradient composition precision is a key factor for the precision of peak retention times and, therefore, peak areas. If this parameter is measured and approaches the limit, the peak retention time precision will also soon exceed the specified limits.

Documentation

The documentation of testing should include

- the description and unique identification of equipment,
- test items,
- acceptance criteria,
- summary of results,
- date when the test was performed and
- names and signatures of persons who performed the tests.

If the tests were performed by a manufacturer's representative, the test report should be signed by the vendor's and the user's representative. The instrument should be labeled with the calibration and qualification status indicating the dates of the last and next calibration and OQ.

A Practical and Economical Approach for Implementation

Several qualification aspects were discussed in the previous chapter. Now it is relevant to give recommendations for practical implementation.

1. Use only one documented procedure for all instrument categories of the same type in your lab, preferably within your company. This significantly reduces the number of documents and time for training of personnel.

2. For all instrument categories in a laboratory (gas chromatographs, for example), use the same OQ procedure and the same test compounds. This makes it easy to compare instruments against each other; new instruments can be compared with existing ones, and it is easier to set specifications for future purchases.

3. For all instruments in a laboratory, use the same acceptance limits, independent of the age, brand and actual performance of the instrument.

4. The procedures and the acceptance limits should be selected so that in normal circumstances all instruments pass the test. Therefore, the instrument with the worst performance will determine the acceptance limits.

5. If there are applications on specific instruments that require more stringent performance limits for specific applications, make an exemption of this instrument, and set the limits to the more stringent value.

6. Define the time distance between two OQs so that the instrument will pass the test with high probability.

7. Define in an SOP the scope of the OQ. For example, should it prove retrospectively and prospectively that the equipment was and will be fit for its intended use, or should this be done just for future use?

8. In case OQ is for future use, plan preventive maintenance before OQ.

9. Always start with the test of the full system (holistic testing). If that test does not meet the criteria, test individual modules to identify the module that caused the problem.

10. Make a technical and operational evaluation on whether to do OQ using your own staff or the vendor's representatives. If the vendor's procedure does not deviate greatly from your expectations, ask if the vendor can make adjustments.

11. Always ask the vendor for help. Even if you decide to do OQ using your own staff, the vendor should still assist you by providing test procedures, certified standards for testing and software for automated testing.

12. Generate a test report that includes a table with test items, your acceptance criteria, actual results and whether or not the test met the criteria. An example is shown in Figure 6.2. Keep this report for regulatory purposes.

HPLC Verification Report

Test method: C\HPCHEM\1\VERIF\Check.M
Data file directory: C\HPCHEM\1\VERIF\Result.D
Original orperator: Dr. Watson

Test Item	User Limit	Actual	Com
DAD noise	$< 5 \times 10^{-5}$ AU	1×10^{-5} AU	Pass
Baseline drift	$< 2 \times 10^{-3}$ AU/h	1.5×10^{-4} AU/h	Pass
DAD WL calibration	±1 nm	±1 nm	Pass
DAD linearity	1.5 AU	2.2 AU	Pass
Pump performance	< 0.3% RSD RT	0.15% RSD RT	Pass
Temp. stability	±0.15°C	±0.15°C	Pass
Precision of peak area	< 0.5% RSD	0.09% RSD	Pass

Verification Test Overall Results   Pass
HP 1100 Series System, Friday, January 16, 1998
Test Engineer
Name: _____ Signature: _____

Figure 6.2. OQ report obtained from the HP ChemStation for the HP1100 Series HPLC.

Summary Recommendations

1. Develop an operating procedure for OQ.

2. If the vendor offers OQ services, make an economic evaluation on whether OQ by the vendor or the user's firm will be more cost-effective.

3. Use generic chemical standards for testing if the equipment will be used for several different applications. Use an application specific standard if the instrument will be used for one application only.

4. If there are multiple instruments of the same category in a lab, use the same procedure and acceptance limits for all instruments.

5. Set the acceptance limit higher than the manufacturer's specification. This may be up to a factor of 5 or 10. For those instruments that require more stringent values to demonstrate their fitness for the intended use, an exception should be made, and the limits should be set to more stringent values.

6. For modular systems, test the system as a whole, not module by module.

7. Set the time intervals between two OQs so that the actual test results in general are at least 30 percent away from the limits.

8. Chemical standards used for instrument calibration or qualification tests should be traceable to national standards. If the system is intended to be used for different analytical methods, a generic chemical standard with known stability should be used. If the system does not perform as expected, individual modules may be recalibrated and interchangeably used to identify the source of the system problem; thus, modular testing is recommended for troubleshooting purposes.

9. Document OQ. If OQ was performed by the vendor, the OQ document should also be signed by the vendor's representative.

7. Performance Qualification and Maintenance

What will be discussed in this chapter?

1. Procedures to ensure ongoing equipment performance
2. Content of an instrument logbook
3. Type and frequency of performance testing
4. Frequency and parameters for suitability testing
5. Development and interpretation of QC charts
6. Procedures for error handling

> Performance qualification (PQ) is the process of demonstrating that an instrument consistently performs according to a specification appropriate for its routine use [15].

The important word here is *consistently*. The test frequency is higher than for OQ. Another difference is that PQ should always be performed under conditions that are the same as, or similar to, those for routine sample analysis. For a chromatograph, this means using the same column, the same analysis conditions and the same or similar test compounds and sample matrices.

As shown in earlier chapters, validation and qualification are not single occurrence events; they should be performed over the entire life of the equipment. During routine use, procedures should exist to demonstrate that the equipment *will continue to do what it purports to do*. However, testing should not be the only activity to ensure ongoing reliable data. Preventive maintenance, ongoing training for new operators and an appropriate error detection system are equally important.

Each laboratory should have a comprehensive QA program that is well understood, accepted and followed by individuals, as well as by laboratory organizations, to prevent, detect and correct problems. The purpose of this program is to ensure that the equipment is running without problems and that analytical results have the highest probability of being of acceptable quality. Ongoing activities may include the following:

1. Preventive instrument maintenance
2. Regular calibration
3. Full or partial OQ checks
4. Daily check of critical performance characteristics, for example, baseline noise of a UV detector if limit of detection is critical
5. Daily system suitability testing
6. Analysis of blanks
7. Duplicate analysis
8. Analysis of QC samples
9. Procedures to detect, record and handle errors and other unforeseen events
10. Changes to the system in a controlled manner
11. Internal audits
12. Participation in proficiency testing schemes
13. Ongoing training programs for new employees

Items 1 to 9 are specific to measurements and are, therefore, covered in this chapter. Internal audits, people qualification and proficiency testing affect many other activities and will be discussed in three separate chapters. The frequency and need for PQ activities should be based on the type of equipment, the instrument's application and previous experience with the equipment and should be documented in in-house procedures.

All calibration and maintenance activities, errors, repairs, performance tests and other events should be recorded in a logbook. This chapter therefore begins with the organization and content of an equipment logbook.

Logbook

For each instrument, a logbook should be prepared for operators and service technicians to record all equipment-related activities in chronological order. The logbook must be readily available near the equipment, for example, in a drawer next to the instrument or attached to the instrument itself. Information in the logbook may include the following:

- Logbook identification (number, valid time range)
- Instrument identification (manufacturer, model name/number, serial number or reference to the IQ document with serial numbers, firmware revision or reference to the IQ document with firmware revisions, date received, service contact)
- Column entry fields for dates, times and events, for example, initial installation and calibration, module and system updates, errors, repairs, performance tests, QC checks, cleaning and (preventive) maintenance, as well as fields for the name and signature of the technician making the entry.

Events that have involved a repair should always include

- the observed symptom,
- what was repaired and
- what was tested after the repair.

Currently, the most convenient format for such a logbook is a bound paper book format or a 2- or 3-ring binder where forms can easily be added. There is also a clear trend toward the use of electronic notebooks. Some instruments even have such notebooks included to enable information to be entered on a local instrument controller or computer. The logbook should be archived together with calibration and analyses data. Figure 7.1 shows an example of an extract from a logbook of an HPLC system.

Logbook ID, valid from:	HPLC 14, valid from June 17, 1997
Name of equipment:	Liquid Chromatograph with ChemStation
Manufacturer:	Hewlett-Packard
Model:	HP1100 Series
Serial numbers:	See IQ document of HPLC System A3
Service contact:	HP Paramus xxxx-xxxxx

Date	Event	Name	Signature
3/7/97	lamp intensity below acceptable limit; exchanged lamp on UV detector, (P/N HP 4523-6784); measured intensity profile; measured baseline noise	K. Weber	*K. Weber*
12/10/97	system suitability tests results showed retention time precision above limit; pump seal changed (P/N HP 1056-5349); performed system suitability test	K. Weber	*K. Weber*
07/11/97	installed new HP DeskJet printer (HP720C) on ChemStation; performed ChemStation performance verification	M. Bauer	*M. Bauer*
04/01/98	exchanged HP1050 UV detector; performed lamp intensity, wavelength accuracy, baseline noise and baseline drift tests; all tests passed specifications (for details see the qualification workbook)	M. Bauer	*M. Bauer*
Page 3			

Figure 7.1. Extract from an HPLC instrument logbook.

Application specific items that are part of the analysis system but that are frequently changed should not be documented in the instrument logbook; they should be recorded on the daily run sheet for sample runs on that particular system. Examples of this are analytical columns and guard columns.

A well-organized logbook can help to identify possible sources of data errors that have occurred at any specific time. It also helps to identify the expected life span of maintenance parts.

The key to success of any logbook is for it to be used by the operators. Availability of the logbook close to the instrument and a clear structure with easy-to-enter fields for entries will help to achieve this.

Maintenance

Operating procedures for maintenance should be in place for every system component that requires periodic calibration and/or preventive maintenance. Preventive maintenance of hardware should be designed to detect problems before they occur. Critical parts should be listed and be available at the user's site. The procedure should describe

- the maintenance to be done,
- when is it to be done,
- what should be tested afterwards and
- the necessary qualifications for the engineer performing the tasks.

The system components should be labeled with the dates of the last and next scheduled maintenance. All maintenance activities should be documented in the instrument's logbook. Suppliers of equipment should provide a list of recommended maintenance activities and documented procedures on how to perform the maintenance. They also should provide a list with recommended test procedures after maintenance activities. Some vendors also offer maintenance contracts with services for preventive maintenance at scheduled time intervals. A set of diagnostic procedures is performed and critical parts are replaced to avoid or identify problems that have not yet reached the point where they may have an impact on the proper functioning of the system.

Traditionally, maintenance parts are replaced on a set time basis. For example, an HPLC pump seal is replaced every six months, a detector's lamp every three months or so. This is neither economical for the laboratory nor environmentally friendly because frequently the parts would not necessarily need to be exchanged at that particular time. It is better to exchange maintenance parts on a usage basis, as implemented on HP's HPLCs through the early maintenance feedback (EMF). An example is shown in Figure 7.2. The user can enter set limits for the lamp, the solvent pumped through and the number of injections. The instruments record the time usage; if the limits are exceeded, the user is informed via the user interface. This allows timely exchange of the maintenance parts before instrument performance drops below the acceptable limit. The elapsed time after which maintenance should be carried out depends on the particular application. For example, the time after which an HPLC pump seal should be exchanged depends on the mobile phase. The lamp life of an HPLC UV detector depends on the level of baseline noise that is still tolerable for a specific application. The best usage time for a specific part and application should be taken from experience.

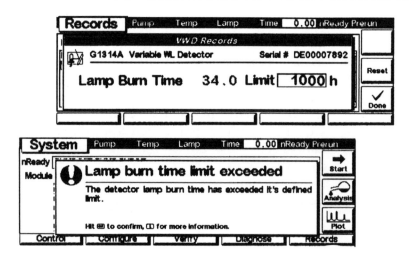

Figure 7.2. HP's early maintenance feedback system (EMF) informs the user when usage limits are reached. Here the burn time and limits for the detector lamp are given.

Calibration

After a certain period of time, operating devices may require recalibration if these are not to impact the performance of an instrument adversely (e.g., the wavelength of a UV-visible detector's optical unit). A calibration program should be in place to recalibrate critical instrument components following documented procedures, with all results recorded in the instrument's logbook. The system components should be labeled with the date of the last and next calibration. The label on the instrument should include the initials of the test engineer, and the calibration report should include his or her printed name and full signature. It is recommended to use forms for instrument calibration, with entry fields for instrument type and serial number, the test frequency, the expected value and acceptance limits and the date and results of actual measurements. An example is shown for an analytical balance in Figure 7.3.

Instrument: Vendor, type
Serial number: 5605AX2
Maximal weight: 110 g
Control weight 1:   10,000 mg   Limit: ±1.0 mg
Control weight 2:   1,000 mg   Limit: ±0.3 mg
Control weight 3:   100 mg   Limit ±0.2 mg
Test frequency: Every day, when used

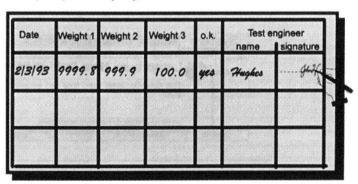

Figure 7.3. Results of calibration are entered on forms.

Performance Testing

The characteristics of equipment alter over time due to contamination and general wear and tear. HPLC UV detector flow cells become contaminated, pump piston seals abrade and UV detector lamps lose intensity. These changes have a direct impact on the performance of analytical hardware and may have a negative effect on the analytical data; therefore, the performance of analytical instruments should be tested during routine use. CITAC's *International Guide to Quality in Analytical Chemistry* [23] specifies the need for performance checks in addition to maintenance and calibration:

> Correct use combined with periodic servicing, cleaning, and calibration will not necessarily ensure an instrument is performing adequately. Where appropriate, periodic performance checks should be carried out (for example, to check the response, stability and linearity of sources, sensors and detectors, the separating efficiency of chromatographic systems, the resolution, alignment and wavelength accuracy of spectrometers etc.).

The performance of equipment should be tested on a frequent basis, for example, daily or each time the instrument is used. The test frequency depends not only on the stability of the equipment but on everything in the system that may contribute to the analysis results. For a liquid chromatograph, this could, for example, be the chromatographic column. The test criteria and frequency should be determined during the development and validation of the analytical method.

In practice, PQ can mean system suitability testing where critical key system performance characteristics are measured and compared to documented, preset limits. Aspects of PQ are often built into analytical methods or procedures. This approach is often called system suitability checking (SSC) and demonstrates that the performance of the measuring procedure (including instrument operating conditions) is appropriate for a particular application. For example, a well-characterized standard may be injected five or six times, and the standard deviation of amounts is subsequently compared against a predefined value. The analysis of QC samples, together with the measurement of certain critical performance characteristics (for example, a detector's baseline noise for trace level analysis) is also suitable for PQ measurements. Table 7.1 includes six steps that can be carried out during preparation and PQ itself.

The user of the equipment carries full responsibility for these activities. The supplier can provide recommendations on what to check, the procedures with test conditions, recommendations for performance limits (acceptance criteria) and recommended actions in case criteria are not met. PQ should follow documented procedures.

Table 7.1. Steps for performance qualification

1. Define the performance criteria and test procedures. These may be derived from OQ tests or from analytical methods or procedures.
2. Select critical parameters. For a chromatography system, these could be
   - a detector's baseline noise,
   - precision of the amounts,
   - precision of retention times,
   - resolution between two peaks,
   - peak width at half height or
   - peak tailing.
3. Select acceptance criteria.
4. Define test intervals, e.g.,
   - every day,
   - every time the system is used or
   - before, between and after a series of runs.
5. Define corrective actions on what to do if the system does not meet the criteria, i.e., if the system is out of specification.
6. Perform tests as specified in (1), at intervals as specified in (4), check the results against the acceptance criteria as specified in (3) and take corrective actions, if necessary, as specified in (5).

Which performance characteristics should be tested and how often? A recommendation on the frequency of performance checks is given in the CITAC guide [23]:

> The frequency of such performance checks will be determined by experience and based on need, type and previous performance of equipment. Intervals between the checks should be shorter than the time the equipment has been found to take to drift outside acceptable limits.

This interpretation means that the frequency of performance checks for a particular instrument depends on the *acceptable limits* specified by the user. The more stringent the limits, the sooner the instrument will drift out of them, thus increasing the frequency of the performance checks. The time interval between checks should be identified by experience and documented for each instrument.

Appendix B of the CITAC guide [23] lists parameters to be checked for chromatographic instruments, including liquid and ion chromatographs; for heating/cooling apparatus, including freeze-dryers, freezers, furnaces, hot-air sterilizers, incubators; for spectrometers, autosamplers, microscopes and electrodes. The frequency of checks for other equipment, including balances, volumetric glassware, hydrometers, barometers, timers and thermometers is also listed.

A good recommendation is to carry out performance checks more frequently for new instruments. If the instrument continually meets the performance specifications, the time interval can always be increased.

System Suitability Testing

The mechanisms proposed to prove that systems perform as expected for their intended use are system suitability tests or the analysis of QC samples by constructing control charts. It is recommended that users perform the checks once a day, or even more frequently, depending on the stability of the system and the number of samples analyzed daily.

System suitability tests have been proposed and defined for chromatographic systems by the USP and other pharmacopeias. Compared to method validation, daily system suitability testing requires fewer individual determinations. A general recommendation is to check those parameters that are critical to analysis accuracy and that may change over a relatively short time. The exact type and frequency of tests should be defined during method validation. As a minimum requirement for compound analysis for chromatographic systems, the USP [35] recommends the following measurements:

- Precision of peak areas (system precision)
- Resolution between two compounds
- Tailing factor

Baseline noise and drift and precision of retention times are other possible parameters necessary, for example, when the detection limit or the stability of retention times is critical to the analysis.

System precision is determined by repeatedly injecting a standard solution and measuring the relative standard deviation of the resulting peak areas or peak heights. For the USP monographs [35], unless otherwise noted, 5 replicate chromatograms are required when the stated relative standard deviation (RSD) is 2 percent or less. For values greater than 2 percent, 6 replicate chromatograms should be used. For bioanalytical samples, the percentage RSD should not exceed 15 percent, except at the limit of detection where it should be less than 20 percent [52].

Quality Control Samples with Control Charts

The analysis of QC samples by constructing control charts has been suggested as a way to incorporate quality checks on results as they are being generated. Such tests can then flag the values that may be erroneous for any of the following reasons:

- Reagents are wrongly mixed.
- Reagents are contaminated.
- GC carrier gas is impure.
- HPLC mobile phase is contaminated.
- Instrument characteristics have changed over time.

ISO 7870

Control charts general guide and introduction

1993

Reference 5

84 *Validation and Qualification in Analytical Laboratories*

For an accurate quality check, QC samples are interspersed among the samples themselves at intervals determined by the total number of samples and the precision and reproducibility of the method (Figure 7.4). The control sample frequency depends mainly on the known stability of the measurement process; a stable process requires only occasional monitoring. The CITAC guide states that 5 percent of sample throughput should consist of QC samples for routine analysis and 20 to 50 percent for more complex procedures.

Control samples should have a high degree of similarity to the actual samples analyzed; otherwise one cannot draw reliable conclusions on the measurement system's performance. Control samples must be so homogeneous and stable that individual increments measured at various times have less variability than the measurement process itself. QC samples are prepared by adding known amounts of analytes to blank specimens. They can be purchased as certified reference materials or may be prepared in-house. QC materials based on environmental matrices, food, serum or urine are commercially available for a variety of analytes. For day-to-day

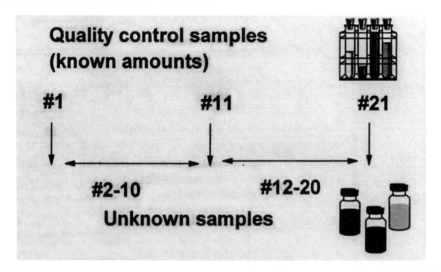

Figure 7.4. For an accurate quality check, QC samples are interspersed among actual samples.

routine analysis, it is recommended to use in-house standards that are checked against a certified reference material. Sufficient quantities should be prepared to enable the same samples to be used over a longer period of time. Their stability over time should be proven, and their accuracy verified, preferably through a comparison with a certified reference material, through interlaboratory tests or by other analysis methods.

The most widely used procedure for the ongoing control of equipment, using QC samples, involves the construction of control charts. These are plots of multiple data points versus the number of measurements from the same QC samples using the same processes. Measured concentrations of a single measurement, or the average of multiple measurements, are plotted on the vertical axis, with the sequence number of the measurement on the horizontal axis. Control charts provide a graphic tool to

- demonstrate statistical control,
- monitor a measurement process,
- diagnose measurement problems and
- document measurement uncertainty.

ISO 8258

Shewhart control charts

1991

Reference 8

Many schemes for the construction of such control charts have been put forward. This book has only limited scope for describing control charts and the statistical theory on which they are based. Details on how to collect data and on how to construct Shewart control charts are described in the ISO Guides 7870 and 8258 [5,8] and in Reference 99.

The most commonly used control charts are X-charts and R-charts, as developed by Dr. Walter Shewhart in 1924 [8]. Both charts are often plotted together as X/R-charts. R-charts plot the range of results obtained from two or more measurements. This shows any change in the dispersion of the process. X-charts either plot single results points for single measurements or the average values from multiple measurements. They consist of a central line representing either the known concentration or the mean of 10 to 20 earlier determinations of the analyte in a control material (QC sample). The standard deviation, determined during method validation, is used to calculate the control lines in the control chart. Control limits define the bounds of virtually all values produced by a system in statistical control.

X-charts (Figure 7.5) often have a center line and two control lines with two pairs of limits: a warning line at $\mu \pm 2\sigma$ and an action line at $\mu \pm 3\sigma$. Statistics predict that 95.45 percent and 99.7 percent of the data will fall within the areas enclosed by the $\pm 2\sigma$ and $\pm 3\sigma$ limits. The center line is either the mean or the true value. In the ideal case, where unbiased methods are used, the center line is the true value. This applies, for example, to precision control charts for standard solutions.

When the process is under statistical control, the day-to-day results are normally distributed about the center line, and 1 out of 20 results is expected to fall between the warning and action lines. No action is required if only one result falls in this area, provided that the next value is inside the warning line. However, if two consecutive values fall between the warning and action lines, then there is evidence of loss of a statistical control. Seven or more consecutive points above the 50 percent confidence limit indicates a tendency for the process to get out of control. More out-of-control situations are shown in Figure 7.6. In these cases, the results should be

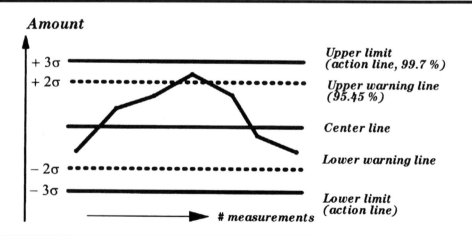

Figure 7.5. X-chart with warning lines and control lines.

a) one value outside the control limit
b) seven consecutive values ascending or descending
c) seven consecutive values above or below the center line
d) two out of three consecutive values outside the warning limits
e) difference between 2 consecutive values >4σ

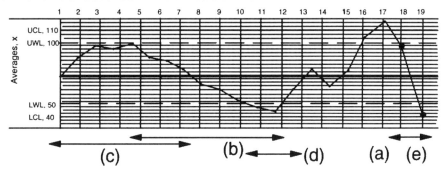

Figure 7.6. Possible out-of-control events.

rejected and the process investigated for its unusual behavior. Further analyses should be suspended until the problem is resolved. Instruments and sampling procedures should be checked for errors.

QC samples may have to be run in duplicate at three concentrations corresponding to the levels below, within and above the analysis range. For methods with linear concentration-response relationships over the full analysis range, two concentrations, one each at the high and low end of the range, are adequate.

An ideal control sample should simulate sample compounds and sample matrices as closely as possible. Other criteria for control samples are

- safe to use for the laboratory staff;
- stable over time;
- long lasting;

- cost-effective; and
- traceable to any national or international standards, if such standards are available.

In routine analytical analysis, the control sample amounts are typically plotted versus the sample number as quality characteristics. This is a useful measurement because it indicates what may come up during sample preparation and measurement. In chromatography, other control parameters may be considered, for example, the resolution between two peaks, the width of a specific peak at the half peak height or the tailing factor (Figure 7.7). Measuring and plotting these parameters gives useful hints when the system approaches the limits of specified ranges, and corrective actions can be initiated before wrong data are measured. For example, if in liquid chromatography the resolution between two peaks drops below a specified limit, or the tailing factor goes above a certain limit, it is most likely that the column needs to be changed.

A documented quality procedure should be in place that provides the operator with step-by-step instructions in the event that the results of one or more QC samples are outside the warning or control line. There are two types of corrective action: immediate on-the-spot and long-term. On-the-spot action is used to correct minor problems, such as the replacement of defective instrument parts, an HPLC UV-visible

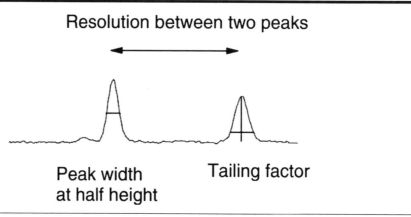

Figure 7.7. Possible quality characteristics in chromatography.

detector lamp, for example. These actions can be performed by a single individual, and analytical methods or procedures do not need to be changed. Long-term action is required when an out-of-control situation (Table 7.2) is caused by a method, an uncommon equipment failure or laboratory environment problem.

For long-term actions, one person is made responsible for investigating the cause, developing and implementing corrective action and verifying that the problem has been solved.

Handling of Defective Instruments

Clear instructions should be available to the operator on actions to take in the event that an instrument breaks down or fails to function properly. Recommendations should be given on when operators should attempt to rectify the problem themselves and when they should call the instrument vendor's service department. In cases of malfunction, it is not sufficient to repair the instrument on-site and then to continue performing analyses. For each instrument, there should be a list of common and uncommon failures, and every problem should be classified in this way (Figure 7.8). Common problems, such as a defective UV-visible detector lamp, require short-term action. The lamp should be replaced and, after a functional test, the instrument can be used for further analyses. The failure, repair and result of the functional test should be entered into the instrument's logbook.

Table 7.2. Possible actions in out-of-control events

- Check materials (reagents, solvents and calibration standards for correct weighing, within specified time for stability, different supplier).
- Check QC sample (correct weighing, storage conditions, within stability time).
- Check if the right method has been used.
- Check instrumentation (hardware, software, correct integration, sufficient separation, sufficient precision).
- Check whether the operator changed.

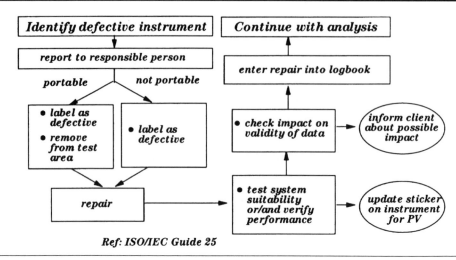

Figure 7.8. Handling of instruments with uncommon failures.

In the case of an uncommon failure that cannot be easily classified and repaired by the operator, several steps are required:

- The problem should be reported to the laboratory supervisor, or to the person responsible for the instrument, who will decide on further action.

- The instrument should be removed from the laboratory and stored in a specified area or, if this is impractical due to its size, it should be clearly labeled as being defective. For example, portable equipment such as pH meters should be removed, while larger equipment such as an HPLC, a GC or an ICP-MS system should be labeled "out of service."

- After repair, correct functioning must be verified, the type and extent of testing depending on the failure and possible impact on the system. Depending on the failure, this may require part or full performance verification (requalification) or only system suitability testing.

- The impact of the defect on previous test results should be examined.

- Clients should be informed about the effect the failure may have had on the validity of their data.

- An entry on the defect, repair and performance verification should be made in the instrument's logbook.

Summary Recommendations

1. Develop an equipment logbook.

2. Develop maintenance procedures (with the help of the vendor).

3. Develop procedures and acceptance limits for performance testing (criteria: regulations, instrument type, application, performance requirements).

4. Develop procedures in case acceptance criteria are not met.

5. Develop procedures in case of equipment failures.

8. Operational Qualification of Software and Computer Systems

What will be discussed in this chapter?

1. How to classify different categories of computer systems in laboratories

2. How to qualify computerized analysis systems

3. How to qualify computer networks

4. How to validate and qualify user-contributed software programs (e.g., Macros)

5. How to qualify existing systems

6. What documentation should be generated

Introduction

The correct functioning of software and computer systems should be verified after installation and before routine use. While in the past regulatory agencies did not pay much attention to software and computer systems, this has recently changed. For example, the OECD consensus paper number 10 [47] requires acceptance testing, which is part of an OQ.

OQ for software and computer systems is more difficult than for hardware for four reasons:

1. It is more difficult to define specifications for software.

2. It is more difficult to define test procedures and acceptance criteria.

3. There are hardly any guidelines available on the OQ of software and computer systems.

4. While equipment hardware performance problems are easily identified, this is not always the case with software. Even though they may be present from the start, they may become evident only after certain combinations of software modules are executed.

Because of these problems, there is even more uncertainty for software and computer systems than for equipment hardware. The basic questions are as follows:

- How much testing is enough?

- Should all functions be tested?

- How do I perform the tests?

- If I have multiple computers with the same configurations, should I repeat all tests for all systems?

Too much testing can become quite expensive, and insufficient testing can be a problem during an audit. For example, the author has seen test protocols of 200 and more pages that users of an off-the-shelf commercial computerized chromatographic system developed over several weeks. Each software function, such as switching the integrator on and off, has been verified as part of an OQ procedure. This is not necessary if the tests have been done and documented at the vendor's site.

This chapter has been dedicated to the uncertainty and the special problems that occur with software and computer systems. However, because the scope of this book covers all aspects of validation, the author cannot furnish enough background to acquire the basics of software and computer validation. This subject matter, with many examples and SOPs, is discussed and documented in a book dedicated to this topic [49].

The type of testing required for the qualification of software and computer systems depends very much on the type and complexity of the software. We can differentiate between three different situations leading into further discussions in this chapter:

Interpharm

Validation of Computerized Analytical Systems

**Ludwig Huber
1995**

Reference 49

1. Vendor-supplied software and computer hardware is an integral part of an analysis system, for example, a computerized spectrographic system where the computer is used for instrument control, data acquisition and data evaluation. Testing of computer functions can be done while processing reference samples.

2. Several computer systems are interconnected to each other and may also be interfaced to analytical systems. Examples are client/servers and laboratory information management systems (LIMS).

3. Software has been developed in the user's laboratory as an add-on to a vendor-supplied software package, e.g., a Macro or a stand-alone software package.

Indeed, in practice, computer systems found in analytical laboratories are combinations of categories 1, 2 and 3. The validation requirements for each category will be discussed separately. If combinations of the categories are used, the validation activities can also be combined. Testing is very different among the categories, but the basic procedure is the same for all three categories.

1. Define the functions.

2. Develop test cases, and define expected results and acceptance criteria.

3. Approve test plan before the tests start.

4. Execute the tests.

5. Compare the results with the expected results and acceptance criteria.

6. Approve and document everything.

Computerized Analyses Systems

Correct functioning of software loaded on a computer system should be checked in the user's laboratory under typical operating conditions. During the equipment hardware test, as described in the previous section, many software functions are executed (Figure 8.1).

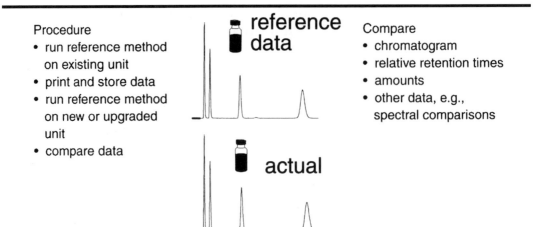

Figure 8.1. Testing of computer hardware and software as part of an integrated chromatographic system.

- Instrument control
- Data acquisition
- Peak integration
- Quantitation
- File storage
- File retrieval
- Printing
- Security control (if implemented)

Therefore, with the successful completion of hardware tests, it can also be assumed that the software operates as intended. Another possibility is to run a well-characterized test sample under normal and stress conditions (e.g., run multiple systems in parallel and compare the newly calculated results with results from previous runs).

There are two situations where software verification, independent of the equipment hardware, may be necessary:

1. Not all critical software functions are executed during the hardware verification (e.g., spectral evaluation).

2. A verification of the software functions is done without a need for equipment testing. This is the case after a change on the computer system, for example, if a new version of the operating system has been installed or if new hardware, such as CD-ROMs, internal memory (RAM) or a hard disk, has been installed on the computer system.

Most software functions of a chromatographic computerized system can also be tested by using well-characterized data files without injecting a test sample. The advantage is that less time is required for the test. The concept has been described in great detail in Reference 49 and is summarized in Table 8.1.

Table 8.1. Qualification process of chromatographic computer systems using data files

Generation of Master Data

1. Generate one or more master chromatograms (the chromatograms should reflect typical samples).
2. If the method uses spectral data, generate master spectra with spectral libraries.
3. Develop integration method, calibration method and procedure for spectral evaluation (e.g., peak purity and/or identity checks).
4. Generate and print out master result(s).
5. Save master chromatograms, master method and results on paper and store them electronically as data files.

Verification

1. Select data file with master chromatogram (and spectra).
2. Select file with master method for data processing (integration, quantitation, spectral evaluation, etc.).
3. Run test manually or automatically. Automation is preferred because it is faster and has less chance for errors to occur.
4. Compare test results with master data. Again, this can be done automatically if such software is built into the system.
5. Print and archive results.

The procedure is very useful after updating the computer system, for example, after adding more internal memory or when changing to a new operating system. The test procedure is very generic and can also be used to test and verify the correct functions of other software packages.

Well-characterized test chromatograms and spectra derived from standards or real samples are stored on disk as a master file. Chromatograms and spectra may be supplied by the vendor as part of the software package. The vendor-supplied chromatograms and spectra are only useful if they reflect the user's way of working; otherwise, test chromatograms and spectra should be recorded by the user. This *master data file* passes through normal data evaluation from spectral evaluation and integration to report generation. Results are stored on the hard disk. Exactly the same results should always be obtained when using the same data file and method for testing purposes. If the chromatographic software is used for different methods, the test should be for different methods. For example, one test can be a setup assay and another can be for impurity tests.

Preferably, tests and documentation of results should be done automatically, always using the same set of test files. In this way, users are encouraged to perform the tests more frequently, and user-specific errors are eliminated. In some cases, vendors provide test files and automated test routines for verification of a computer system's performance in the user's laboratory. Needless to say, the correct functioning of this software should also be verified. This can easily be done by changing the method or data file and rerunning the test. The report should indicate an error. If automated verification software is not available, the execution of the tests, the verification of actual results with prerecorded results and documentation can be done manually.

Successful execution of this procedure ensures that

- the actual version of the application software works correctly for the tested functions;
- executed program and data files are loaded correctly on the hard disk;
- the actual computer hardware is compatible with the software; and

- the actual version of the operating system and user interface software is compatible with the application software.

Computer Network

For a networked computer system, OQ can mean, for example, checking security and verifying correct communication between the computers and peripherals (Figure 8.2). Data sets should be developed and input on one part of the network. The output at some other part should be compared with the input. For example, if a server is used to secure and archive data from a chromatographic data station, results should be printed on

- the chromatographic data system and
- the server after storage and retrieval of the files.

The results should be compared, either manually or automatically.

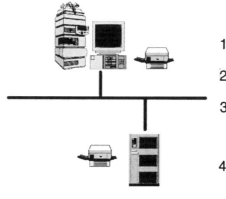

Integrated analytical system

1. Qualify integrated system
2. Print data set on integrated system
3. Send data to network, store and retrive data, print data on network printer
4. Compare data with (2)

Network server

Figure 8.2. Qualification of a network server for data storage and printing.

Existing Systems and Systems Without Vendor Validation

Existing computer systems in laboratories may require retrospective evaluation and qualification if their initial validation was not formally documented. Typical questions are as follows:

- Do I have to validate existing systems?

- Should the same validation criteria and procedures be applied for existing systems as for new systems?

- What if I cannot get any documented evidence from the vendor about validation during development?

- What type of testing do I have to do?

- How can I use the test data that I have collected in the past?

There is no doubt that existing systems should be validated if data generated on the system are critical and if the system will be used to generate such data; however, it is difficult to use the same validation criteria for an older computer system as those used for a new one. The software might not have been developed in accordance with the most recent product life cycle guidelines, and full documentation may not have been archived.

Fortunately, existing computer systems have an advantage not shared by new systems—the experience gained over time. The validation process can take advantage of this wealth of historical experience by reviewing the quality of analytical results obtained from computerized systems. Such a review may provide sufficient evidence that the system has done and is still doing what it is supposed to do. In this case, retrospective evaluation and validation are just a matter of documenting what already has been done in the past.

The validation of existing systems takes time, and it is quite obvious that not all existing systems can be validated at one particular time; it may not even be necessary to validate all systems in a laboratory. Therefore, the validation process should follow a multistep plan:

1. Identify all analytical systems with computers in a laboratory.

2. Identify those systems that need to be validated.

3. Develop a validation schedule for those systems that need to be validated.

4. Implement the schedule.

Before a decision is made to qualify an existing system retrospectively, serious thought should be given whether to purchase a new system or to update the current one. Important criteria to consider are the anticipated costs of a validation versus a new system and an estimation on how successful the validation will be. The latter can be estimated by looking at the history of the computer system and checking for regular maintenance, calibrations and performance checks and trouble-free operation over a long period.

Once the decision has been made to qualify the system, a plan and documentation should be prepared. Ideally, the same documentation should be available for existing systems, as described in the previous chapter, as for new systems; every attempt should be made to acquire this information.

The qualification protocol for an existing system should include a list of missing documentation usually required for validation. The protocol should also provide an explanation as to why the documentation is missing. In many cases, the qualification may have been performed, but the relevant data was not documented. In other cases, the data may have been retained, but proper authorization signatures were not obtained. The validation plan should also contain a contingency plan that describes what should be done if the previously generated data are deemed to be incorrect (e.g., who should be notified).

After evaluation and qualification, the following documentation should be available:

- The qualification plan and protocol
- A description of the system hardware and software
- Historical logs of hardware with system failure reports, maintenance logs and records, as well as calibration records

- Test data demonstrating that the system does what it purports to do (this can be system suitability test results or well-documented QC charts)

- Procedures and schedules for preventive maintenance and ongoing performance testing (e.g., regular system suitability tests or the analysis of QC samples)

- A plan for error recording, reporting and remedial action

A similar qualification procedure is recommended for new systems if the vendor will not, or cannot, provide evidence of development validation. However, this qualification requires more intensive testing because the system has provided no previous data. If the analysts are familiar with the technique, and well-defined reference samples are available, testing the system as a whole using the reference samples with the anticipated operating ranges can be adequate. Users should compare the results with results obtained from other instruments currently in use.

If the analysts have little or no experience with the analysis technique, use the modular test approach to examine each system and software program module by module, checking for correct instrument control, peak integration, compound quantitation, and data storage and retrieval. When testing individual items, the author recommends developing templates that include a cover page with information about test environments, instrument configuration and software revision. The templates should include entry fields for observations of actual results made during testing. Simple pass-fail indications are insufficient.

Tests should be included that check the system's error handling capabilities. The system should recognize and display any wrong entry, such as flow rates greater or smaller than the operational range. Another simple test could check how the program responds when alphabetic data are input to entry fields that are designed to accept numeric data. The tests should also check system boundary conditions. To test these conditions, input data that are slightly greater or less than the operational limits. For example, if the operational limit of a gas chromatograph's oven is 400°C, try entering values of 399°C and 401°C.

User-Contributed Software (e.g., Macros)

Application software developed by the user should be fully validated and documented by the user. Such software may be a stand-alone software package (e.g., for statistical data evaluation), or it may be an extension to purchased standard software (e.g., a Macro to enhance functionality). The development and validation of such software should follow a documented procedure, and the source code should be available. The effort involved in validation depends very much on the size and complexity of the program. The development of large programs should follow the software development life cycle and can take several weeks or months. Validation can take several weeks, and the documentation will be extensive. On the other hand, the validation of smaller programs can be done in a few hours, and the documentation may be only a few pages. The development, validation and documentation of such small programs requires, at minimum, the following steps:

1. Describe the problem, how the problem is solved currently and how the newly developed program will solve it.

2. Identify responsibilities for development, test and approvals.

3. Describe the task and the system requirements (hardware, system software, standard software).

4. Describe the program in terms of the functions it will perform.

5. Document formulas and algorithms used within the code.

6. Write and document the code in such a way that it can be understood by other people whose knowledge and experience are similar to the programmer's. Print the code.

7. Develop test cases and data sets with known inputs and outputs. Include test cases with normal data across the operating range, some at the boundary and some unusual cases with incorrect inputs. The results should be calculated by the new program and also by using alternative methods. The development of an automated test

procedure that can be executed as often as possible is recommended. Test procedures and results should be documented, reviewed and signed off.

8. Develop user documentation with information on how to install, test and operate the program.

9. Describe and implement procedures for data backup and security routines for limited access to authorized people.

10. Develop a procedure to authorize, test, document and approve any changes to the software and documentation.

For combined systems, vendor-updated software revisions may be critical, especially if the updated version supplied by the vendor will have an effect on the interface between the vendor's and the user's software (e.g., if the meaning of a Macro command has been changed). The user should obtain information from the vendor on how the updated version may affect the interface. The user should also test his or her software after it has been integrated into the vendor's updated standard software. More details about SOPs for developing and validating simple, as well as complex, application software developed in the user's laboratory are found elsewhere [49].

Implementation and Documentation

Some important points should be considered for the implementation of OQ of software and computer systems:

1. For complex systems, a validation team should be formed consisting of analysis experts from the laboratories affected, computer experts from IT departments and validation experts.

2. An overall validation plan should be developed that describes the purpose of the system, including subsystems, responsible persons, test philosophy and a schedule for testing. OQ should be part of this plan.

3. The intended use and functions of the network and all subsystems should be defined. For subsystems and core functions of the network, the vendor should provide a list with detailed functional specifications. From these

specifications, the user can derive the functions the systems will employ in the user's laboratory.

4. For networked systems, test cases should be developed for each subsystem, and each subsystem should be validated. Again, the vendor should provide validated software to execute these tests automatically. An example was given above for chromatographic data systems.

5. A test plan with test cases and acceptance criteria should be developed and approved before the tests start.

6. For multiuser systems, some tests should be done while the maximum number of users are operating the system.

7. All or at least some tests should be done under the maximum data flow.

8. When there is a change to the system, the validation team should evaluate the possible impact of the change on other parts of the system. Based on this evaluation, a test plan should be developed that executes either all or part of the tests as specified in step 3.

At the end of OQ, documentation should be available or developed that includes a validation protocol with

- the description, intended use and unique identification of equipment;
- functional specifications;
- test protocols to include test items, acceptance criteria, actual test results, date and time when tests have been performed and a list of the names of people who performed the tests as well as signatures; and
- summary of results and a statement on the validation status.

Summary Recommendations

1. Classify the computer system.
2. For complex systems, establish a validation team.
3. Define intended functions, tests and acceptance criteria.

4. Execute and document tests.

5. Develop and implement a procedure for software developed in the user's laboratory.

6. For existing systems, evaluate and qualify the system based on past experience.

7. If there is no evidence of sufficient testing, the system should be tested like a new system.

9. Validation of Analytical Methods

What will be discussed in this chapter?

1. The parameters that should be validated
2. Difference between USP and ICH procedures
3. How to develop and implement a strategy for method validation
4. How to validate standard methods
5. How to implement validated methods in routine use
6. When and how a method should be revalidated

Introduction

Method validation is the process used to confirm that the analytical procedure employed for a specific test is suitable for its intended use. Results from method validation can be used to judge the quality, reliability and consistency of analytical results; it is an integral part of any good analytical practice.

Analytical methods need to be validated or revalidated

- before their introduction into routine use;
- whenever the conditions change for which the method has been validated (e.g., an instrument with different characteristics or samples with a different matrix); and
- whenever the method is changed and the change is outside the original scope of the method.

Method validation has received considerable attention in the literature and from industrial committees and regulatory agencies. The *Guidance on the Interpretation of the EN 45000 Series of Standards* and *ISO/IEC Guide 25* includes a chapter on the validation of methods [21] with a list of nine validation parameters. The ICH [18] has developed a consensus text on the validation of analytical procedures. The document includes definitions for eight validation characteristics and appendices with more detailed methodology [19].

The U.S. EPA prepared a guidance for methods development and validation for the Resource Conservation and Recovery Act (RCRA) [3]. The AOAC, the EPA and other scientific organizations provide methods that are validated through multilaboratory studies.

The U.S. FDA has proposed guidelines on submitting samples and analytical data for methods validation [2, 53]. The USP has published specific guidelines for method validation for compound evaluation [20]. USP 23 defines eight steps for validation:

1. Accuracy
2. Precision
3. Specificity
4. Limit of detection
5. Limit of quantitation
6. Linearity and range
7. Ruggedness
8. Robustness

The U.S. FDA has proposed adding Section 211.222 on "methods validation" to the cGMP regulations [54]. This would require the manufacturer to establish and document the accuracy, sensitivity, specificity, reproducibility and any other attribute necessary to validate test methods. Validation would be required to meet the existing requirements for laboratory records provided in Sec. 211.194 (a). These requirements include a statement of each method used in testing the sample to meet proper standards of accuracy and reliability, as applied to the tested product.

There are no official guidelines referring to biological fluids. The pharmaceutical industry uses the methodology published in literature [52, 62]. The most comprehensive document is the conference report of the 1990 Washington conference: *Analytical Methods Validation: Bioavailability, Bioequivalence and Pharmacokinetic Studies*, which was sponsored by, among others, the American Association of Pharmaceutical Scientists (AAPS), the AOAC and the U.S. FDA [52]. The report presents guiding principles for validating studies of both human and animal subjects that may be referred to in developing future formal guidelines.

Representatives of the pharmaceutical and chemical industry have published papers on the validation of analytical methods. Hokanson [55, 56] applied the life cycle approach, developed for computerized systems, to the validation and revalidation of methods. Green [57] gave a practical guide for analytical method validation, with a description of a set of minimum requirements for a method. Renger and his colleagues [58] described the validation of a specific analytical procedure for the analysis of theophylline in a tablet using high-performance thin layer chromatography (HPTLC). The validation procedure in this particular article is based on requirements for EU multistate registration. Wegscheider [59] has published procedures for method validation with a special focus on calibration, recovery experiments, method comparison and investigation of ruggedness. Seno et al. [60] have described how analytical methods are validated in a Japanese QC laboratory. The AOAC [4] has developed a Peer-Verified Methods validation program with detailed guidelines on exactly which parameters should be validated. Winslow and Meyer [61] recommend the definition and application of a master plan for validating analytical methods.

This chapter gives a review and a strategy for the validation of analytical methods for both methods developed in-house as well as standard methods, and a recommendation on the documentation that should be produced during, and on completion of, method validation.

Strategy for the Validation of Methods

The validity of a specific method should be demonstrated in laboratory experiments using samples or standards that are similar to unknown samples analyzed routinely. The preparation and execution should follow a validation protocol, preferably written in a step-by-step instruction format. Possible steps for a complete method validation are listed in Table 9.1. This proposed procedure assumes that the instrument has been selected and the method has been developed. It meets criteria such as ease of use; ability to be automated and to be controlled by computer systems; costs per analysis; sample throughput; turnaround time; and environmental, health and safety requirements.

Successful acceptance of the validation parameters and performance criteria, by all parties involved, requires the cooperative efforts of several departments, including analytical

Table 9.1. Steps in method validation

1. Develop a validation protocol, an operating procedure or a validation master plan for the validation.
2. Define the application, purpose and scope of the method.
3. Define the performance parameters and acceptance criteria.
4. Define validation experiments.
5. Verify relevant performance characteristics of equipment.
6. Qualify materials, e.g., standards and reagents for purity, accurate amounts and sufficient stability.
7. Perform prevalidation experiments.
8. Adjust method parameters and/or acceptance criteria if necessary.
9. Perform full internal (and external) validation experiments.
10. Develop SOPs for executing the method in routine analysis.
11. Define criteria for revalidation.
12. Define type and frequency of system suitability tests and/or analytical quality control (AQC) checks for routine analysis.
13. Document validation experiments and results in the validation report.

development, QC, regulatory affairs and the individuals requiring the analytical data. The operating procedure or the Validation Master Plan (VMP) should clearly define the roles and responsibilities of each department involved in the validation of analytical methods.

The scope of the method and its validation criteria should be defined early in the process. These include the following questions:

- What analytes should be detected?
- What are the expected concentration levels?
- What are the sample matrices?
- Are there interfering substances expected, and, if so, should they be detected and quantified?
- Are there any specific legislative or regulatory requirements?
- Should information be qualitative or quantitative?
- What are the required detection and quantitation limits?
- What is the expected concentration range?
- What precision and accuracy is expected?
- How robust should the method be?
- Which type of equipment should be used? Is the method for one specific instrument, or should it be used by all instruments of the same type?
- Will the method be used in one specific laboratory or should it be applicable in all laboratories?
- What skills do the anticipated users of the method have?

The method's performance characteristics should be based on the intended use of the method. It is not always necessary to validate all analytical parameters that are available for a specific technique. For example, if the method is to be used for qualitative trace level analysis, there is no need to test and validate the method's limit of quantitation, or the linearity, over the full dynamic range of the equipment. Initial parameters should be chosen according to the analyst's experience and best judgment. Final parameters should be agreed between the lab or analytical chemist performing the validation and the lab or individual applying the method and users of the data to be generated by the method. Table 9.2 gives examples of which parameters might be tested for a particular analysis task.

The scope of the method should also include the different types of equipment and the locations where the method will be run. For example, if the method is to be run on a specific instrument in a specific laboratory, there is no need to use instruments from other vendors or to include other laboratories in the validation experiments. In this way, the experiments can be limited to what is really necessary.

The validation experiments should be carried out by an experienced analyst to avoid errors due to inexperience. The analyst should be very well versed in the technique and operation of the instrument. Before an instrument is used to validate a method, its performance specifications should be verified using generic chemical standards. Satisfactory results for a method can be obtained only with equipment that is

Table 9.2. Validation parameters for different analysis tasks

	Major Compounds—Quantitative	Major Compounds and Traces—Quantitative	Traces—Qualitative	Traces—Quantitative
Limit of detection	no	no	yes	no
Limit of quantitation	no	yes	no	yes
Linearity	yes	yes	no	yes
Range	yes	yes	no	no
Precision	yes	yes	no	yes
Accuracy	yes	yes	no	yes
Specificity	yes	yes	yes	yes
Ruggedness	yes	yes	no	yes

performing well. Special attention should be paid to those equipment characteristics that are critical for the method. For example, if detection limit is critical for a specific method, the instrument's specification for baseline noise and, for certain detectors, the response to specified compounds should be verified.

Any chemicals used to determine critical validation parameters, such as reagents and reference standards, should be

- available in sufficient quantities,
- accurately identified,
- sufficiently stable and
- checked for exact composition and purity.

Any other materials and consumables, for example, chromatographic columns, should be new. This ensures that one set of consumables can be used for most experiments and avoids unpleasant surprises during method validation.

Operators should be sufficiently familiar with the technique and equipment. This will allow them to identify and diagnose unforeseen problems more easily and to run the entire process more efficiently.

If there is little or no information on the method's performance characteristics, it is recommended to prove the suitability of the method for its intended use in initial experiments. These studies should include the approximate precision, working range and detection limits. If the preliminary validation data appear to be inappropriate, the method itself, the equipment, the analysis technique or the acceptance limits should be changed. Method development and validation are, therefore, an iterative process. For example, in liquid chromatography, selectivity is achieved through the selection of mobile phase composition. For quantitative measurements, the resolution factor between two peaks should be 2.5 or higher. If this value is not achieved, the mobile phase composition needs further optimization. The influence of operating parameters on the performance of the method should be assessed at this stage if this was not done during development and optimization of the method.

There are no official guidelines on the correct sequence of validation experiments, and the optimal sequence may depend on the method itself. Based on the author's experience, for a liquid chromatographic method, the following sequence has proven to be useful:

1. Selectivity of standards (optimizing separation and detection of standard mixtures if selectivity is insufficient)
2. Linearity, limit of quantitation, limit of detection, range
3. Repeatability (short-term precision) of retention times and peak areas
4. Intermediate precision
5. Selectivity with real samples
6. Trueness/accuracy at different concentrations
7. Ruggedness (interlaboratory studies)

The more time-consuming experiments, such as accuracy and ruggedness, are included toward the end. Some of the parameters, as listed under (2) to (6), can be measured in combined experiments. For example, when the precision of peak areas is measured over the full concentration range, the data can be used to validate the linearity.

During method validation, the parameters, acceptance limits and frequency of ongoing system suitability tests or QC checks should be defined. Criteria should be defined to indicate when the method and system are beyond statistical control. The aim is to optimize these experiments so that, with a minimum number of control analyses, the method and the complete analytical system will provide long-term results to meet the objectives defined in the scope of the method.

Once the method has been developed and validated, a validation report should be prepared that includes the following:

- Objective and scope of the method (applicability, type).
- Summary of methodology.
- Type of compounds and matrix.
- All chemicals, reagents, reference standards, QC samples with purity, grade, their source or detailed instructions on their preparation.
- Procedures for quality checks of standards and chemicals used.

- Safety precautions.
- A plan and procedure for method implementation from the method development lab to routine analysis.
- Method parameters.
- Critical parameters taken from robustness testing.
- Listing of equipment and its functional and performance requirements, e.g., cell dimensions, baseline noise and column temperature range. For complex equipment, a picture or schematic diagram may be useful.
- Detailed conditions on how the experiments were conducted, including sample preparation. The report must be detailed enough to ensure that it can be reproduced by a competent technician with comparable equipment.
- Statistical procedures and representative calculations.
- Procedures for QC in routine analyses, e.g., system suitability tests.
- Representative plots, e.g., chromatograms, spectra and calibration curves.
- Method acceptance limit performance data.
- The expected uncertainty of measurement results.
- Criteria for revalidation.
- The person who developed and validated the method.
- References (if any).
- Summary and conclusions.
- Approval with names, titles, date and signature of those responsible for the review and approval of the analytical test procedure.

Validation of Standard Methods

A laboratory applying a specific method should have documented evidence that the method has been appropriately validated. This holds for methods developed in-house, as well as for standard methods, for example, those developed by organizations such as the EPA, American Society for Testing and Materials (ASTM), ISO or the USP.

A number of questions usually arises about the validation of standard methods: Firstly, should these methods be revalidated in the user's laboratory and, if so, should method revalidation cover all experiments, as performed during initial validation? Secondly, which documentation should be available or developed in-house for standard methods? Official guidelines and regulations are not explicit about validating standard methods. Only the CITAG guide [23] includes a short paragraph that reads as follows:

> The validation of standard or collaboratively tested methods should not be taken for granted, no matter how impeccable the method's pedigree—the laboratory should satisfy itself that the degree of validation of a particular method is adequate for the required purpose, and that the laboratory is itself able to match any stated performance data.

There are two important requirements in this excerpt:

1. The standard's method validation data must exactly meet the laboratory's method requirements.

2. The laboratory must be able to match the performance data as described in the standard.

This chapter elaborates on what these statements mean in practice, and it gives a strategy for validating standard methods.

Like the validation of methods developed in-house, the evaluation and validation of standard methods should also follow a documented process that is usually the validation plan. Results should be documented in the validation protocol. Both documents will be the major source for the validation report.

An example of a step-by-step plan for the evaluation and validation of standard methods is shown as a flow diagram in Figure 9.1. As a first step, the scope of the method, as applied in the user's laboratory, should be defined. This should be done independently of what is written in the standard method and should include information such as

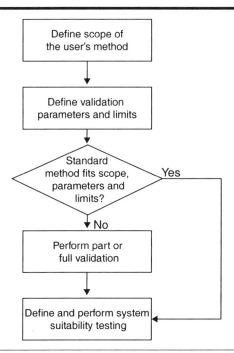

Figure 9.1. Workflow for evaluation and validation of standard methods.

- the type of compounds to be analyzed,
- matrices,
- the type of information required (qualitative or quantitative),
- detection and quantitation limits,
- range,
- precision and accuracy as specified by the client of the analytical data and
- the type of equipment—its location and environmental conditions.

As a second step, the method's performance requirements should be defined in considerable detail, again irrespective of what has been validated in the standard method. General guidelines on validation criteria for different measurement objectives and procedures for their evaluation are discussed later in this chapter.

The results of these steps lead to the experiments that are required for adequate method validation and to the minimal acceptance criteria necessary to prove that the method is suitable for its intended use. Third, required experiments and expected results should be compared with what is written in the standard method.

In particular, the standard method should be checked for the following items:

1. Have the reported validation results been obtained from the complete procedure or from just a part of it? Sometimes the validation data from the published method have been obtained from the chromatographic analysis but have not included sample preparation steps. The diagram in Figure 9.2 can be used for this check. A complete validation of the analytical procedure should include the entire process from sampling, sample preparation, analysis, calibration and data evaluation to reporting.

2. Has the same matrix been used?

3. Did the validation experiments cover the complete concentration range as intended for the method in the user's

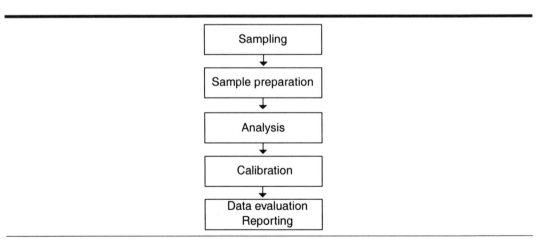

Figure 9.2. Steps for validating complete analytical procedures. Standard methods should be checked if all steps are included in the validation data.

laboratory? If so, has the method's performance been checked at the different concentration ranges?

4. Has the same equipment (brand, model) been used as available in the user's laboratory, and, if not, was the scope of standard method regarding this item broad enough to include the user's equipment? This question is very important for a gradient HPLC analysis, where the HPLC's delay volume can significantly influence the method's selectivity.

5. Have performance characteristics, e.g., the limit of quantitation, been checked in compliance with the most recent guidelines, as required for the user's laboratory (e.g., the ICH guidelines [19] for pharmaceutical laboratories)? If not, does the test procedure have equivalency to the guideline?

If either the scope, the validation parameters or the validation results do not meet the user's requirements, adequate validation experiments should be defined, developed and carried out. The extent of these experiments depends on the overlap of the user requirements with the scope and results, as described in the standard method. If there is no overlap, a complete validation should be carried out. In the case of a complete overlap, validation experiments may not be necessary.

If method validation experiments are unnecessary, the user should prove the suitability of the method in his or her laboratory. This evidence should confirm that the user's equipment, the people, the reagents and the environment are qualified to perform the analysis. The experiments may be an extract of the full method validation and should focus on the critical items of the method. Guidelines for these tests should have been developed during method development. If not, they should be developed and carried out at this stage. Typical experiments may include precision of amounts and limits of quantitation.

The validation report should include a reference to the standard method.

Validation of Nonroutine Methods

Frequently, a specific method is used for only a few sample analyses. The question should be raised as to whether this method also needs to be validated using the same criteria as recommended for routine analysis. In this case, the validation may take much more time than the sample analysis and may be considered inefficient, because the cost per sample will increase significantly. The answer is quite simple: Any analysis is worthwhile only if the data are sufficiently accurate; otherwise, sample analysis is pointless. The suitability of an analysis method for its intended use is a prerequisite to obtaining accurate data; therefore, only validated methods should be used to acquire meaningful data. However, depending on the situation, the validation efforts can be reduced for nonroutine methods. The CITAG guide [23] includes a chapter on how to treat nonroutine methods. The recommendation is to reduce the validation cost by using generic methods, for example, methods that are broadly applicable. A generic method could, for example, be based on capillary gas chromatography or on reversed phase gradient HPLC. With little or no modification, the method can be applied to a large number of samples. The performance parameters should have been validated on typical samples characterized by sample matrix, compound types and concentration range.

If, for example, a new compound with a similar structure in the same matrix is to be analyzed, the validation will require only a few key experiments. The documentation of such generic methods should be designed to easily accommodate small changes relating to individual steps, such as sample preparation, sample analysis or data evaluation.

The method's operating procedure should define the checks that need to be carried out for a novel analyte in order to establish that the analysis is valid. Detailed documentation of all experimental parameters is important to ensure that the work can be repeated in precisely the same manner at any later date.

Quality Control Plan

For any method that will be used for routine analysis, a QC plan should be developed. This plan should ensure that the method, together with the equipment, delivers consistently accurate results. The plan may include recommendations for the following:

1. Selection, handling and testing of QC standards

2. Type and frequency of equipment checks and calibrations (for example, should the wavelength accuracy and the baseline noise of an HPLC UV detector be checked after each sample analysis, or on a daily or weekly basis?)

3. Type and frequency of system suitability testing (for example, at which point during the sequence system should suitability standards be analyzed?)

4. Type and frequency of QC samples (for example, should a QC sample be analyzed after 1, 5, 20 or 50 unknown samples, and should there be single or duplicate QC sample analysis, or should this be run at one or several concentrations?)

5. Acceptance criteria for equipment checks, system suitability tests and QC sample analysis

6. Action plan in case criteria 2, 3 and/or 4 are not met.

Implementation to Routine Analysis

In many cases, methods are developed and validated in service laboratories that are specialized in this task. When the method is transferred to the routine analytical laboratory, care should be taken that the method and its critical parameters are well understood by the workers in the departments who apply the method. A detailed validation protocol, a documented procedure for method implementation and good communication between the development and operation departments are equally important. If the method is used by a number of departments, it is recommended to verify method validation parameters and to test the applicability and usability of the method in a couple of these departments before it is distributed to other departments. In this way, problems can be identified and corrected before the method is distributed to a larger audience. If the method is intended to be used by just one or two departments, an analyst from the development department should assist the users of the method during initial operation. Users of the method should be encouraged to give constant feedback on the applicability and usability of the method to the development department. The latter should correct problems if any arise.

Revalidation

Operating ranges should be defined for each method, either based on experience with similar methods or else investigated during method development. These ranges should be verified during method validation in robustness studies and should be part of the method characteristics. Availability of such operating ranges makes it easier to decide when a method should be revalidated. A revalidation is necessary whenever a method is changed, and the new parameter lies outside the operating range. If, for example, the operating range of the column temperature has been specified to be between 30 and 40°C, the method should be revalidated if, for whatever reason, the new operating parameter is 41°C.

Revalidation is also required if the scope of the method has been changed or extended, for example, if the sample matrix changes or if operating conditions change. Furthermore, revalidation is necessary if the intention is to use instruments with different characteristics, and these new characteristics have not been covered by the initial validation. For example, an HPLC method may have been developed and validated on a pump with a delay volume of 5 mL, but the new pump has a delay volume of only 0.5 mL.

Part or full revalidation may also be considered if system suitability tests, or the results of QC sample analysis, lie outside preset acceptance criteria and where the source of the error cannot be traced back to the instruments or any other cause.

Whenever there is a change that may require part or full revalidation, the change should follow a documented change control system. A flow diagram of such a process is documented in Figure 9.3. The change should be defined, authorized for implementation and documented. Possible changes may include

- new samples with new compounds or new matrices,
- new analysts with different skills,
- new instruments with different characteristics,
- new location with different environmental conditions,
- new chemicals and/or reference standards and
- modification of analytical parameters.

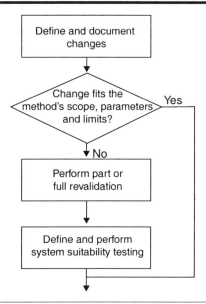

Figure 9.3. Workflow for revalidation.

An evaluation should determine whether the change is within the scope of the method. If so, no revalidation is required. If the change lies outside the scope, the parameters for revalidation should be defined. After the validation experiments, the system suitability test parameters should be investigated and redefined, if necessary.

Parameters for Method Validation

The parameters for method validation have been defined in different working groups of national and international committees and are described in the literature. Unfortunately, some of the definitions vary between the different organizations. An attempt at harmonization was made for pharmaceutical applications through the ICH [18, 19], where representatives from the industry and regulatory agencies from the United States, Europe and Japan defined parameters, requirements and, to some extent, methodology for analytical methods validation. The parameters, as defined by the ICH and by other organizations and authors, are summarized in Table 9.3 and are described in brief in the following paragraphs.

Table 9.3. Possible analytical parameters for method validation

- Specificity (1,2)
- Selectivity
- Precision (1,2)
    - repeatability (1)
    - intermediate precision (1)
    - reproducibility (3)
- Accuracy (1,2)
- Trueness
- Bias
- Linearity (1,2)
- Range (1,2)
- Limit of detection (1,2)
- Limit of quantitation (1,2)
- Robustness (2,3)
- Ruggedness (2)

(1) Included in ICH publications
(2) Included in the USP
(3) Terminology included in ICH publication but not part of required parameters

Selectivity/Specificity

The terms *selectivity* and *specificity* are often used interchangeably. A detailed discussion of this term, as defined by different organizations, has been presented by Vessmann [63]. He particularly pointed out the difference between the definitions of specificity given by IUPAC/WELAC and the ICH.

Although it is not consistent with the ICH, the term *specific* generally refers to a method that produces a response for a single analyte only, while the term *selective* refers to a method that provides responses for a number of chemical entities that may or may not be distinguished from each other. If the response is distinguished from all other responses, the method is said to be selective. Since there are very few methods that respond to only one analyte, the term *selectivity* is usually more appropriate. The USP monograph [20] defines the selectivity of an analytical method as its ability to measure accurately an analyte in the presence of interference, such as synthetic precursors, excipients, enantiomers and known (or likely) degradation products that may be expected to be present in the sample matrix. Selectivity in liquid chromatography is obtained by choosing optimal columns and setting chromatographic conditions, such as mobile phase composition, column temperature and detector wavelength. Besides chromatographic separation, the sample preparation step can also be optimized for best selectivity.

It is a difficult task in chromatography to ascertain whether the peaks within a sample chromatogram are pure or consist of more than one compound. Therefore, the analyst should know how many compounds are in the sample or whether procedures for detecting impure peaks should be used.

While in the past chromatographic parameters such as mobile phase composition or the column were modified, now the application of spectroscopic detectors coupled on-line to the chromatograph is being used. UV/visible diode-array detectors and mass spectrometers acquire spectra on-line throughout the entire chromatogram. The spectra acquired during the elution of a peak are normalized and overlaid for graphical presentation. If the normalized spectra are different, the peak consists of at least two compounds.

The principles of diode-array detection in HPLC and their application and limitations with regard to peak purity are described in the literature [64–66]. Examples of pure and impure HPLC peaks are shown in Figure 9.4. While the chromatographic signal indicates no impurities in either peak, the spectral evaluation identifies the peak on the left as impure. The level of impurities that can be detected with this method depends on the spectral difference, on the detector's performance and on the software algorithm. Under ideal conditions, peak impurities of 0.05 to 0.1 percent can be detected.

Selectivity studies should also assess interferences that may be caused by the matrix, e.g., urine, blood, soil, water or food. Optimized sample preparation can eliminate most of the matrix components. The absence of matrix interferences for a

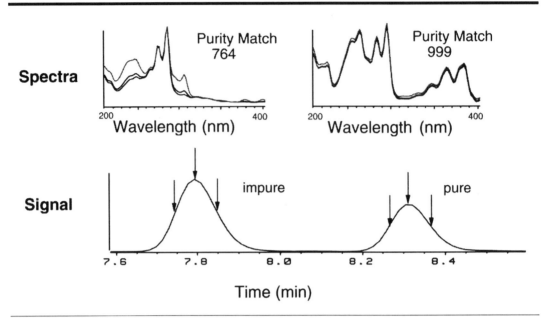

Figure 9.4. Examples of pure and impure HPLC peaks. The chromatographic signal does not indicate any impurity in either peak. Spectral evaluation, however, identifies the peak on the left as impure.

quantitative method should be demonstrated by the analysis of at least five independent sources of control matrix.

Precision and Reproducibility

The precision of a method (Table 9.4) is the extent to which the individual test results of multiple injections of a series of standards agree. The measured standard deviation can be subdivided into 3 categories: repeatability, intermediate precision and reproducibility [18, 19]. Repeatability is obtained when the analysis is carried out in a laboratory by an operator using a piece of equipment over a relatively short time span. At least 6 determinations of 3 different matrices at 2 or 3 different concentrations should be performed, and the RSD calculated.

The ICH [19] requires precision from at least 6 replications to be measured at 100 percent of the test target concentration or from at least 9 replications covering the complete specified range. For example, the results can be obtained at 3 concentrations

Table 9.4. Analyte concentration versus precision (Ref. 4)

Analyte %	Analyte Ratio	Unit	RSD (%)
100	1	100%	1.3
10	10^{-1}	10%	2.8
1	10^{-2}	1%	2.7
0.1	10^{-3}	0.1%	3.7
0.01	10^{-4}	100 ppm	5.3
0.001	10^{-5}	10 ppm	7.3
0.0001	10^{-6}	1 ppm	11
0.00001	10^{-7}	100 ppb	15
0.000001	10^{-8}	10 ppb	21
0.0000001	10^{-9}	1 ppb	30

with 3 injections at each concentration. The Center of Drug Evaluation and Research (CDER, Rockville, Md., USA) recommends a minimum of 10 injections with a relative standard deviation of less than 1 percent RSD [3].

The acceptance criteria for precision depend very much on the type of analysis. Pharmaceutical QC precision of greater than 1 percent RSD is easily achieved for compound analysis, but the precision for biological samples is more like 15 percent at the concentration limits and 10 percent at other concentration levels. For environmental and food samples, precision is largely dependent on the sample matrix, the concentration of the analyte, the performance of the equipment and the analysis technique. It can vary between 2 percent and more than 20 percent.

The AOAC manual for the Peer-Verified Methods program [4] includes a table with estimated precision data as a function of analyte concentration (Table 9.4).

Intermediate precision is a term that has been defined by ICH [18] as the long-term variability of the measurement process. It is determined by comparing the results of a method run within a single laboratory over a number of weeks. A method's intermediate precision may reflect discrepancies in results obtained

- from different operators,
- from inconsistent working practice (thoroughness) of the same operator,
- from different instruments,
- with standards and reagents from different suppliers,
- with columns from different batches or
- a combination of these.

The objective of intermediate precision validation is to verify that in the same laboratory the method will provide the same results once the development phase is over.

Reproducibility (Table 9.5), as defined by the ICH [18, 19], represents the precision obtained between different laboratories. The objective is to verify that the method will provide the same results in different laboratories. The reproducibility of an analytical method is determined by analyzing aliquots from homogeneous lots in different laboratories with different analysts, and by using operational and environmental conditions that may differ from, but are still within, the specified parameters of the method (interlaboratory tests).

Table 9.5. Typical variations affecting a method's reproducibility

- Differences in room temperature and humidity
- Operators with different experience and thoroughness
- Equipment with different characteristics, e.g., delay volume of an HPLC system
- Variations in material and instrument conditions, e.g., in HPLC mobile phase composition, pH, flow rate
- Variation in experimental details not specified by the method
- Equipment and consumables of different ages
- Columns from different suppliers or different batches
- Solvents, reagents and other material with varying quality

Validation of reproducibility is important if the method is to be used in different laboratories.

Table 9.6 summarizes factors that should be the same, or different, for precision, intermediate precision and reproducibility.

Accuracy and Recovery

The accuracy of an analytical method is the extent to which test results generated by the method and the true value agree. Accuracy can also be described as the closeness of agreement between the value that is adopted, either as a conventional, true or accepted reference value, and the value found.

The true value for accuracy assessment can be obtained in several ways. One alternative is to compare the results of the method with results from an established reference method. This approach assumes that the uncertainty of the reference method is known. Secondly, accuracy can be assessed by analyzing a sample with known concentrations (e.g., a control

Table 9.6. Variables for measurements of precision, intermediate precision and reproducibility

	Precision	Intermediate Precision	Reproducibility
Instrument	same	different	different
Batches of accessories, e.g., chromatographic columns	same	different	different
Operator	same	different	different
Sample matrices	same	different	different
Concentration	same	different	different
Batches of material, e.g., reagents	same	different	different
Environmental conditions, e.g., temperature, humidity	same	different	different
Laboratory	same	same	different

sample or certified reference material) and comparing the measured value with the true value as supplied with the material. If certified reference materials or control samples are not available, a blank sample matrix of interest can be spiked with a known concentration by weight or volume. After extraction of the analyte from the matrix and injection into the analytical instrument, its recovery can be determined by comparing the response of the extract with the response of the reference material dissolved in a pure solvent. Because this accuracy assessment measures the effectiveness of sample preparation, care should be taken to mimic the actual sample preparation as closely as possible. If validated correctly, the recovery factor determined for different concentrations can be used to correct the final results.

The concentration should cover the range of concern and should include concentrations close to the quantitation limit, one in the middle of the range and one at the high end of the calibration curve. Another approach is to use the critical decision value as the concentration point that must be the point of greatest accuracy.

The expected recovery (Table 9.7) depends on the sample matrix, the sample processing procedure and the analyte concentration. The AOAC manual for the Peer-Verified Methods program [4] includes a table with estimated recovery data as a function analyte concentration.

The ICH document on validation methodology recommends accuracy to be assessed using a minimum of nine determinations over a minimum of three concentration levels covering the specified range (e.g., three concentrations/three replicates each). Accuracy should be reported as percent recovery by the assay of known added amount of analyte in the sample or as the difference between the mean and the accepted true value, together with the confidence intervals.

Linearity and Calibration Curve

The linearity of an analytical method is its ability to elicit test results that are directly proportional to the concentration of analytes in samples within a given range or proportional by means of well-defined mathematical transformations. Linearity may be demonstrated directly on the test substance (by dilution of a standard stock solution) and/or by using separate

Table 9.7. Analyte recovery at different concentrations (Ref. 4)

Analyte Ingred. (%)	Analyte ratio	Unit	Mean recovery (%)
100	1	100%	98–102
≥10	10^{-1}	10%	98–102
≥1	10^{-2}	1%	97–103
≥0.1	10^{-3}	0.1%	95–105
0.01	10^{-4}	100 ppm	90–107
0.001	10^{-5}	10 ppm	80–110
0.0001	10^{-6}	1 ppm	80–110
0.00001	10^{-7}	100 ppb	80–110
0.000001	10^{-8}	10 ppb	60–115
0.0000001	10^{-9}	1 ppb	40–120

weighings of synthetic mixtures of the test product components, using the proposed procedure.

Linearity is determined by a series of 3 to 6 injections of 5 or more standards whose concentrations span 80–120 percent of the expected concentration range. The response should be directly proportional to the concentrations of the analytes or proportional by means of a well-defined mathematical calculation. A linear regression equation applied to the results should have an intercept not significantly different from 0. If a significant nonzero intercept is obtained, it should be demonstrated that this has no effect on the accuracy of the method.

Frequently, the linearity is evaluated graphically, in addition to or as an alternative to mathematical evaluation. The evaluation is made by visually inspecting a plot of signal height or peak area as a function of analyte concentration. Because deviations from linearity are sometimes difficult to detect, two additional graphical procedures can be used. The first is to plot the deviations from the regression line versus the concentration or versus the logarithm of the concentration, if the concentration range covers several decades. For linear ranges, the deviations should be equally distributed between positive and negative values.

Another approach is to divide signal data by their respective concentrations, yielding the relative responses. A graph is plotted with the relative responses on the y-axis and the corresponding concentrations on the x-axis, on a log scale. The obtained line should be horizontal over the full linear range. At higher concentrations, there will typically be a negative deviation from linearity. Parallel horizontal lines are drawn on the graph corresponding to, for example, 95 percent and 105 percent of the horizontal line. The method is linear up to the point where the plotted relative response line intersects the 95 percent line. Figure 9.5 shows a comparison of the two graphical evaluations on a sample of caffeine using HPLC.

The ICH recommends, for accuracy reporting, the linearity curve's correlation coefficient, y-intercept, slope of the regression line and residual sum of squares. A plot of the data should be included in the report. In addition, an analysis of the deviation of the actual data points from the regression line may also be helpful for evaluating linearity. Some analytical procedures, such as immunoassays, do not demonstrate linearity after any transformation. In this case, the analytical response should be described by an appropriate function of the concentration (amount) of an analyte in a sample. In order to establish linearity, a minimum of five concentrations is recommended. Other approaches should be justified.

Range

The range of an analytical method is the interval between the upper and lower levels (including these levels) that have been demonstrated to be determined with precision, accuracy and linearity using the method as written. The range is normally expressed in the same units as the test results (e.g., percentage, parts per million) obtained by the analytical method.

For assay tests, the ICH [19] requires the minimum specified range to be 80 to 120 percent of the test concentration, and for the determination of an impurity, the range to extend from the limit of quantitation, or from 50 percent of the specification of each impurity, whichever is greater, to 120 percent of the specification.

Limit of Detection

The limit of detection is the point at which a measured value is larger than the uncertainty associated with it. It is the low-

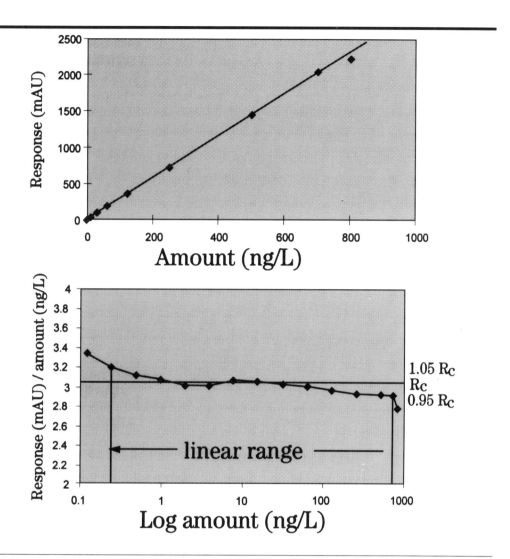

Figure 9.5. Graphical presentations of a linearity plot of a caffeine sample using HPLC. Plotting the sensitivity (response/amount) gives clear indication of the linear range. Plotting the amount on a logarithmic scale has a significant advantage for wide linear ranges. R_c = Line of constant response.

est concentration of analyte in a sample that can be detected but not necessarily quantified. The limit of detection is frequently confused with the sensitivity of the method. The sensitivity of an analytical method is the capability of the method to discriminate small differences in concentration or mass of the test analyte. In practical terms, sensitivity is the slope of the calibration curve that is obtained by plotting the response against the analyte concentration or mass.

In chromatography, the detection limit is the injected amount that results in a peak with a height at least two or three times as high as the baseline noise level. Besides this signal/noise method, the ICH [19] describes three more methods:

1. *Visual inspection:* The detection limit is determined by the analysis of samples with known concentrations of analyte and by establishing the minimum level at which the analyte can be reliably detected.

2. *Standard deviation of the response based on the standard deviation of the blank:* Measurement of the magnitude of analytical background response is performed by analyzing an appropriate number of blank samples and calculating the standard deviation of these responses.

3. *Standard deviation of the response based on the slope of the calibration curve:* A specific calibration curve is studied using samples containing an analyte in the range of the limit of detection. The residual standard deviation of a regression line, or the standard deviation of y-intercepts of regression lines, may be used as the standard deviation.

Limit of Quantitation

The limit of quantitation is the minimum injected amount that produces quantitative measurements in the target matrix with acceptable precision in chromatography, typically requiring peak heights 10 to 20 times higher than the baseline noise.

If the required precision of the method at the limit of quantitation has been specified, the EURACHEM [21] (Figure 9.8) approach can be used. A number of samples with decreasing amounts of the analyte are injected six times. The calculated RSD percent of the precision is plotted against the analyte amount. The amount that corresponds to the previously defined required precision is equal to the limit of quantitation. It is important to use not only pure standards for this test but also spiked matrices that closely represent the unknown samples.

For the limit of detection, the ICH [19] recommends, in addition to the procedures as described above, the visual inspection and the standard deviation of the response and the slope of the calibration curve.

Any results of limits of detection and quantitation measurements must be verified by experimental tests with samples containing the analytes at levels across the two regions. It is equally important to assess other method validation parameters, such as precision, reproducibility and accuracy, close to the limits of detection and quantitation. Figure 9.6 illustrates the limit of quantitation (along with the limit of detection, range and linearity). Figure 9.7 illustrates both the limit of detection and the limit of quantitation.

Ruggedness

Ruggedness is not addressed in the ICH documents [18, 19]. Its definition has been replaced by reproducibility, which has the same meaning as ruggedness, defined by the USP as the degree of reproducibility of results obtained under a variety of conditions, such as different laboratories, analysts, instruments, environmental conditions, operators and materials. Ruggedness is a measure of reproducibility of test results under normal, expected operational conditions from laboratory to laboratory and from analyst to analyst. Ruggedness is

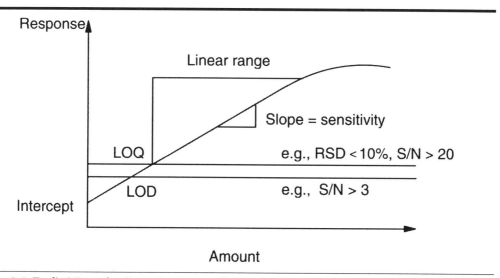

Figure 9.6. Definitions for linearity, range, limit of quantitation (LOQ) and limit of detection (LOD).

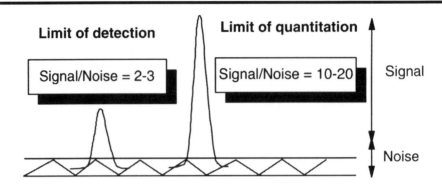

Figure 9.7. Limit of detection and limit of quantitation via signal to noise.

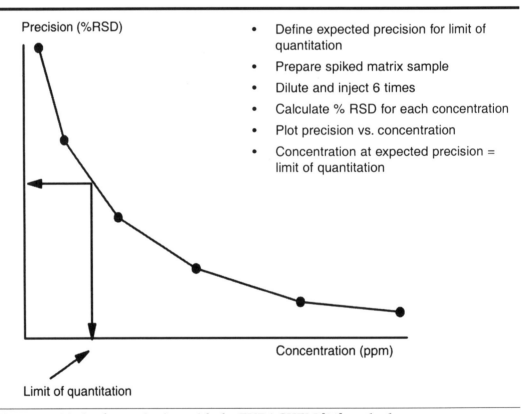

- Define expected precision for limit of quantitation
- Prepare spiked matrix sample
- Dilute and inject 6 times
- Calculate % RSD for each concentration
- Plot precision vs. concentration
- Concentration at expected precision = limit of quantitation

Figure 9.8. Limit of quantitation with the EURACHEM [21] method.

determined by the analysis of aliquots from homogeneous lots in different laboratories.

Robustness

Robustness tests examine the effect that operational parameters have on the analysis results. For the determination of a method's robustness, a number of method parameters, for example, pH, flow rate, column temperature, injection volume, detection wavelength or mobile phase composition, are varied within a realistic range, and the quantitative influence of the variables is determined. If the influence of the parameter is within a previously specified tolerance, the parameter is said to be within the method's robustness range.

Obtaining data on these effects helps to assess whether a method needs to be revalidated when one or more parameters are changed, for example, to compensate for column performance over time. In the ICH document [19], it is recommended to consider the evaluation of a method's robustness during the development phase, and any results that are critical for the method should be documented. This is not, however, required as part of a registration.

Stability

Many solutes readily decompose prior to chromatographic investigations, for example, during the preparation of the sample solutions, extraction, cleanup, phase transfer or storage of prepared vials (in refrigerators or in an automatic sampler). Under these circumstances, method development should investigate the stability of the analytes and standards.

The term *system stability* has been defined as the stability of the samples being analyzed in a sample solution. It is a measure of the bias in assay results generated during a preselected time interval, for example, every hour up to 46 hours, using a single solution (Figure 9.9) [94]. System stability should be determined by replicate analysis of the sample solution. System stability is considered appropriate when the RSD, calculated on the assay results obtained at different time intervals, does not exceed more than 20 percent of the corresponding value of the system precision. If, on plotting the assay results as a function of time, the value is higher, the

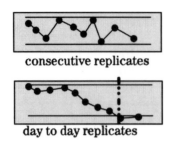

Figure 9.9. Schematics of stability testing.

maximum duration of the usability of the sample solution can be calculated [94].

The effect of long-term storage and freeze-thaw cycles can be investigated by analyzing a spiked sample immediately after preparation and on subsequent days of the anticipated storage period. A minimum of two cycles at two concentrations should be studied in duplicate. If the integrity of the drug is affected by freezing and thawing, spiked samples should be stored in individual containers, and appropriate caution should be employed for the study of samples.

Which Parameters Should Be Included in Method Validation?

For an efficient validation process, it is of utmost importance to specify the right validation parameters and acceptance criteria. The more parameters, the more time it will take to validate. The more stringent the specifications or acceptance limits, the more often the equipment has to be recalibrated, and probably also requalified, to meet the higher specifications at any one time. It is not always essential to validate every analytical performance parameter, but it is necessary to define which ones are required. This decision should be based on business, regulatory and/or accreditation requirements.

1. For contract analyses: What does the client request?
2. For regulatory submission: What do the regulations or guidelines require?
3. For laboratory accreditation: What do the standard and relevant guidelines recommend?

The validation parameters depend on the analytical task and the scope of the method. For example, both the USP [20] and the ICH [18] contain chapters on validation procedures for different analytical tasks, both of which are included to provide some ideas on what type of validations are required for different tasks (see Tables 9.8 and 9.9). For example, according to the ICH, accuracy, any type of precision and limits of detection and quantitation are not required if the analytical task is identification. For assays in USP category 1, the major component or active ingredient to be measured is normally present at high concentrations; therefore, validation of limits of detection and quantitation is not necessary.

Because the type of analysis and the information that should be obtained from a sample have so much influence on the

Table 9.8. ICH validation characteristics

Analytical Task	Identification	Impurity Testing		Assay
		Quantitative	Limit Tests	
Accuracy	no	yes	no	yes
Precision				
Repeatability	no	yes	no	yes
Interm. Precision	no	yes	no	yes
Reproducibility	no	yes	no	yes
Specificity	yes	yes	yes	yes
Limit of detection	no	no	yes	no
Limit of quantitation	no	yes	no	no
Linearity	no	yes	no	yes
Range	no	yes	no	yes

validation, the objective and scope of the method should always be defined as the first step of any method validation.

Summary Recommendations

1. Develop an operating procedure for method validation.
2. Define intended use of the method and performance criteria.
3. Check all equipment and material for performance and quality.
4. Perform validation experiments.
5. For standard methods: check scope of the standard with your own requirements.
6. For nonroutine methods: develop and use generic methods and customize them for specific nonroutine tasks.

Table 9.9. USP validation characteristics

Analytical Task	Assay Category 1	Assay Category 2		Assay Category 3
		Quantitative	Limit Tests	
Accuracy	yes	yes	*	*
Precision	yes	yes	no	yes
Specificity	yes	yes	yes	*
Limit of detection	no	no	yes	*
Limit of quantitation	no	yes	no	*
Linearity	yes	yes	no	*
Range	yes	yes	*	*
Ruggedness	yes	yes	yes	*

* May be required depending on the nature of the specific test.

10. Data Validation and Evaluation of Uncertainty

What will be discussed in this chapter?

1. How to validate data
2. How to identify the contribution of different effects to overall uncertainty
3. How to estimate uncertainty
4. How to report data

Data validation is the process by which data are filtered and accepted or rejected based on defined procedures. It is also the final step before release. The following points should be considered:

- SOPs should exist for the definition of raw data, data entry, security and review.

- The accuracy of critical data should be verified, irrespective of whether the data were entered manually or were transferred electronically from an analytical instrument.

- Checks, preferably performed automatically, should be built into any routine method to identify errors. Requirements for a validity check of data include well-maintained instruments, documented measurement processes and statistically supported limits of uncertainty.

- Final results should be traceable back to the individual who entered the data, or, in cases where data are acquired on-line from an analytical instrument, the instrument should be identified. In the latter case, it is recommended to store the instrument serial number, method parameters and instrument conditions together with the raw data.

- Any failure or unforeseen event that has occurred with the instrument should be recorded automatically in a logbook and stored together with the raw data. The impact of the error on the data should be evaluated and suitable action taken.

- If changes have been made to any data, the original raw data should not be obscured by these changes. The person who made the change must be identified, and the reason for the change should be given together with the date.

- Quantitative data reports should include a statement on the measurement uncertainty. This is the estimate attached to a measurement characterizing the range of values within which the true value is purported to lie (ISO/DIS 3354-1).

This chapter discusses techniques for data validation. It also describes procedures to measure uncertainty and recommends the contents for an analysis report.

Validation of Data

Data should be validated by a qualified and authorized person following an SOP. A prerequisite for accurate data analysis is that the instrument is functioning properly. Preventive maintenance, together with regular calibration and performance verification, facilitates the instrument's ability to generate accurate data.

Checks should be made for

- proper sample identification,
- transmittal errors,
- plausibility and
- consistency.

Techniques used to accomplish this include

- comparisons with similar data,
- checks for plausibility of values (Figure 10.1) with respect to specified limits,

Figure 10.1. Data should be checked for plausibility and approved or rejected. SOPs should be available for the definition and review of raw data.

- regression analysis and
- tests for outliers.

Checks, preferably automated, should be built into any routine method to identify errors. In chromatography and capillary electrophoresis, peak shape, resolution, peak identification and integration marks should be checked to ensure that peaks are suitable for quantitative analysis.

Two frequent points of discussion are how and how much data should be validated. The answer depends on the analysis task, the analysis method and the probability of obtaining any incorrect data. For example, a soil sample is analyzed for organic compounds using HPLC and UV detection. If the analytes are expected to be present close to the detection limits and could possibly interfere chromatographically with the chemical matrix, there is a high risk of wrong identification, integration and quantitation. In this case, it is recommended to inspect every chromatogram visually and to reintegrate if necessary. A 100 percent check of chromatograms is not worth doing, however, for well-defined samples with only a few, well-separated peaks and where the expected amounts are far above the limit of quantitation. Both a good understanding of the analysis task and knowledge of the measurement process, together with a realistic notion of anticipated problems supported by statistical data, are the basis for a sound scientific judgment of the extent of data validation.

Reporting Data

The type of information that is reported depends very much on the individual situation.

Paragraph 5.4.3 of EN 45001 [36] and Paragraph 3.9.2 of the revised ISO Guide 25 [37] give very specific guidelines on the minimal contents of a report. A corresponding template is shown in Figure 10.2. The report should describe

- identification of sample and test items,
- methods for sampling and analysis,
- test results and the expected uncertainty and
- the name of the person responsible for the content.

While the headings should be standardized as much as possible, the presentation of the actual test results should be specifically designed for each type of test and should be easy to understand for the reader.

ISO Guide 25 suggests additional information on a case-by-case basis:

- For nonstandard methods: a brief description of the method
- Deviations from, additions to or exclusions from the test method
- For reports containing the results of sampling: unambiguous identification of substance, matrix, material or product sampled; location of sampling; details of any environmental conditions during sampling that may affect the interpretation of the test results; and a reference to the sampling method.

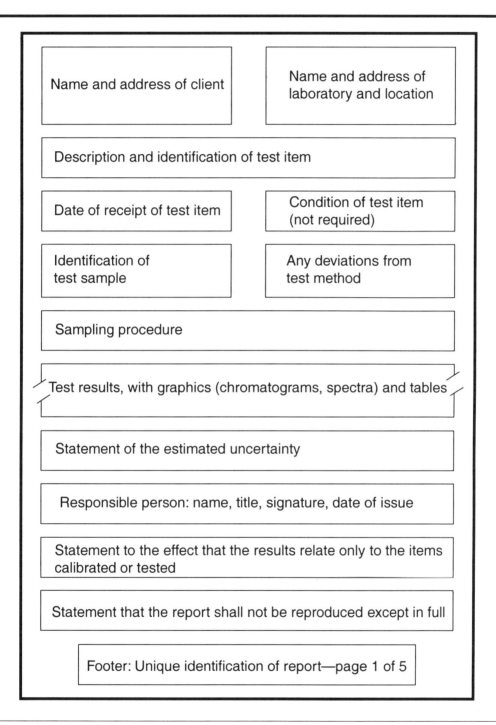

Figure 10.2. Sample for report template with the 14 report items as specified in EN 45001 and ISO Guide 25.

Measurement and Reporting of Uncertainty

Every measurement has an uncertainty associated with it that results from errors arising at the various stages of sampling, sample injection, measurement and data evaluation. In other words, whenever any quantitative measurement is performed, the value obtained is only an approximation of the true value. Users of the measurement data should have an idea of how much the reported result may deviate from the true value. In practice, all accreditation standards and quality guidelines, such as ISO Guide 25, EN 45001 and the CITAC guide, recommend the results of quantitative measurement to be reported as both a single value and together with the possible deviation from the true value. This is the measurement uncertainty. This is logical for any report with quantitative results. It is, for example, of no use if a report on a food sample states 0.1 percent of compound X, and the user of the data is still unsure whether this could be 0.05 or 0.5 percent. An uncertainty statement provides the user with information on the measurement tolerances and the limits within which the true value of the measurement, such as analyte concentration, is supposed to lie. Frequently, the analyst can make a good estimate of the level of uncertainty; the client or user of the data, however, cannot.

In Paragraph 3.9.2.2, the revised ISO Guide 25 [67] contains a passage about reporting measurement uncertainty:

> The report shall include a statement on uncertainty of the test results (this information need only appear in test reports where it is relevant to the validity or application of the test results, where a client's instruction so requires or where uncertainty affects compliance to a specification limit.

Information on uncertainty is of particular importance if a specification limit is to be verified and reported. For example, if, according to a purchasing agreement, a product can only be released if compound X is below 0.5 percent, the test report may not contain a statement about compliance if the measurement results extended by the measurement uncertainty is above 0.5 percent. The revised ISO Guide 25 provides a clear statement on this in paragraph 3.9.2.5:

When parameter(s) are claimed to be within specified tolerance the measurement value(s) extended by the estimated uncertainty of measurement shall fall within the specification limit.

While it is clearly understood that the measurement tolerances should be known and also reported if the client requests these data, there is dissatisfaction among chemists with the word *uncertainty* itself. "Uncertainty" is a negative word and is usually associated with "doubt." This could cast doubt not only on the result but also on the measurement technique, the equipment used, the instrument operator or even the laboratory. Using more positive-sounding words, such as analysis confidence, would create a more positive approach. However, the word *uncertainty* is now well established and should be used whether we like it or not.

ISO has published a *Guide to the Expression of Uncertainty in Measurement* [100]. It establishes general rules for evaluating and expressing uncertainty in measurement across a broad spectrum of measurements.

EURACHEM has produced an excellent document containing many more details on how the concepts of the ISO guide may be applied in chemical measurement. The whole process is schematically shown in Figure 10.3. The basic ideas are explained in this chapter, but for more detailed information, readers of the book are encouraged to study the EURACHEM document [22].

The concept of evaluating uncertainty is fairly straightforward. It requires a detailed knowledge of the nature of the measurand and of the measurement method, rather than an in-depth understanding of statistics. The following steps are recommended:

1. Develop the specifications by writing a clear statement of exactly what is to be measured and the relationship between this and the parameters on which it depends. For example, if the measurement temperature has an influence on the result, the measurement temperature should also be defined.

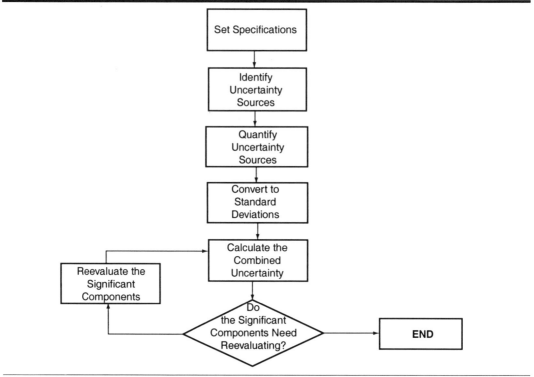

Figure 10.3. Workflow to estimate uncertainty (Ref. 21).

2. Develop a workflow diagram for the entire sampling, sample preparation, calibration, measurement, data evaluation and data transcription process.

3. Identify and list sources of uncertainty for each part of the process or each parameter. Possible sources for errors may be derived from nonrepresentative sampling, operator bias, a wrongly calibrated instrument, lack of ideal measurement conditions, chemicals with impurities and errors in data evaluation.

4. Estimate and document the size of each uncertainty, for example, as standard deviations or as RSDs. These data should be gathered from a series of measurements. Where experimental evaluation is impossible or impractical, the individual contributions should be estimated from whatever sources are available. Sources for this kind of

estimation can be found in the supplier's information or in the results of interlaboratory studies or proficiency testing. The procedures and thoughts behind the way the contributions have been measured or estimated should be documented.

5. Combine separate contributions in order to give an overall value. For example, where individual sources of uncertainty are independent, the overall uncertainty can be calculated as a multiple of the sum of squared contributing uncertainty components, all expressed as standard deviations. Computer software or spreadsheet programs can help to automate this calculation.

The whole procedure should be documented in such a way that sufficient information is available to allow the result to be reevaluated if new information or data become available. A complete documentation should include

- a description of the methods used to calculate the measurement result and its uncertainty from the experimental measurements,
- the values and sources of all corrections and
- a list of all components of uncertainty with full documentation on how each of these was evaluated.

For routine sample analysis, the uncertainty measurement may take place at the end of method validation, before a validated method is used in a laboratory and at discrete intervals. For nonroutine methods, adequate investigation of similar analysis procedures may be sufficient.

Reference 22 includes many practical examples with data from different analyses, as well as formulas for evaluating, calculating and reporting standard and expanded uncertainty. Reports of sample analysis should include an uncertainty number, which is typically expressed as:

$$\text{Result} = x \pm u \text{ (units)}$$
$$\text{or}$$
$$\text{Result} = x \text{ (units)}$$
$$\text{Uncertainty} = u \text{ (units)}$$

Summary Recommendations

1. Develop a general procedure for data validation, traceability and security.

2. Identify the specific data validation need for each analysis type.

3. Develop a general template for reporting. Using this template as a basis, agree with the user of the data on the specific content of the report.

4. Evaluate and report the uncertainty of the overall sampling and analysis procedure. Evaluation can be done during and after method validation and at discrete intervals.

11. (Certified) Reference Standards

What will be discussed in this chapter?

1. Terminology of (certified) reference material; standard reference material; primary, secondary and working standards

2. Regulatory and standard requirements

3. Official reference material programs

4. Requirements for (certified) reference material

5. Requirements for traceability to national or international standards

6. Availability of (certified) reference material

7. How to prepare homemade reference material and working standards

Introduction

ISO/IEC
Guide 35

Certification of
Reference
Materials

*general and
statistical
principles*

1989

Reference 11

The goal of any analytical measurement is to obtain accurate, reliable and consistent data. Prerequisites for achieving accurate results in analytical laboratories are correct sampling, correct weighing of the sample and standards, well-maintained and calibrated equipment, qualified operators, validated methods and procedures for data validation. Most important is the use of accurate standards or (certified) reference materials. No matter how skilled the analysts are or how sophisticated and automated the equipment, if the calibration of the system is incorrect, the analytical result will always be wrong.

Even though the chief role of reference materials is to ensure accuracy for a specific method, there is another, equally important use of such materials: They enable the laboratory and a specific user to verify the performance of equipment, procedures and operators at any time.

Agreement in analysis results with the certified value proves that not only the method is right, but also the equipment and the chemicals used for sample preparation are right and that the operator did a good job. The laboratory can conclude that the data generated for this particular procedure are correct. All laboratories obtaining the same results are "intercalibrated" and in line with the technically competent organization that certified the material. Any disagreement between the certified value and the value determined by the laboratory indicates a problem with the analysis, which then requires a thorough follow-up.

Users may encounter several problems with (certified) reference standards:

- As there are many compounds, it may be difficult to order reference materials for all compounds.

- Even if the compound may be available, the sample matrix may be different from the matrix of the reference materials.

- The concentration may differ.

- Chemical standards may have a limited lifetime.

- Traceability is not always possible.

Because of the importance of reference materials in the overall qualification process and the many problems analysts have with them, this chapter is dedicated to this topic. Frequently asked questions are as follows:

- What is the difference between a reference material and a certified reference material?

- When do I need certified reference materials?

- What do regulations and standard guidelines say?

- Do I need traceability to national or other standards?

- What do I do in case there are no certified standards available?
- How do I prepare working standards in my lab?
- How can I ensure the quality of the reference samples?

Applications of (Certified) Standards

A certified reference material serves multiple purposes in a laboratory.

1. Method validation and revalidation, for example, to validate a method's accuracy, linearity, limit of detection and limit of quantitation.
2. To demonstrate equivalency of a method developed in-house with a standard method.
3. To transfer analytical methods to other laboratories. Correct results with reference standards prove correct functioning in the new environment.
4. To calibrate equipment when the final determinations are based on the measurement of a signal that must be correlated with the concentration of the analyte in the unknown sample. Examples are chromatographic and spectrometric equipment.
5. OQ of analytical equipment, for example, to check the linearity of a detector.
6. To control the overall performance of an analytical procedure, for example, when using QC samples.
7. To check the proficiency of a new person in the lab. Successfully running a reference sample proves the person's qualification to run this type of analysis.
8. Interlaboratory tests to assess either the performance of a method or the proficiency of a laboratory.

Types of Material and Definitions

Different types of reference materials, certified reference materials, standard reference materials and external and internal

> **ISO/IEC Guide 33**
>
> **Uses of Certified Reference Materials**
>
> **1989**
>
> Reference 10

reference materials have been defined by ISO/IEC and the NBS (U.S. National Bureau of Standards).

Reference Material (RM): A material or substance one or more properties of which are sufficiently well established to be used for the calibration of an apparatus, the assessment of measurement method, or for assigning values to materials (ISO IEC Guide 30 - 1992, 2.1) [68].

Certified Reference Material (CRM): A reference material, accompanied by a certificate, one or more of whose property values are certified by a procedure, which establishes its traceability to an accurate realization of the unit in which the property values are expressed, and for which each certified value is accompanied by an uncertainty at a stated level of confidence (ISO IEC Guide 30 - 1992, 2.2) [68].

The terms *reference material* and *certified reference material* have been defined by ISO. When comparing the definitions, it appears that there is more accurate information with a traceability chain in a well-characterized and accurate primary standard, which, for example, may be a national standard. The value and price of certified reference materials are substantially higher than for reference materials.

Certified reference materials are referred to as standard reference material by the NBS, now the National Institute of Standards and Technology (NIST) in the United States.

Standard Reference Material (SRM): Certified reference material issued by the National Bureau of Standards (NBS) [69]

External Reference Material (ERM): Provided by someone other than the National Bureau of Standards (NBS) [69]

Internal Reference Material (IRM): Developed by a laboratory for its own use; also called laboratory standard material (LSM) or working standards (WS)

The Analytical Chemistry Section of IUPAC defined primary, working and secondary standards as follows:

Primary Standard: a commercially available substance of purity 100 ± 0.02% (Purity 99.98 + %)

Working Standard: a commercially available substance of purity 100 ± 0.05% (Purity 99.95 + %)

Secondary Standard: a substance of lower purity which can be standardized against a primary grade standard

Reference materials or certified reference materials can be available as

- pure solutions for single component calibration,
- mixtures in solutions for multicomponent calibration and
- solids with single or multiple components and a matrix as close as possible to the matrix of the unknown sample used as QC samples for long-term performance of a procedure.

Regulatory and Standard Requirements

Because of the importance of reference materials in laboratories, they are subject to the regulations and quality standards related to chemical laboratories. The main requirements are as follows:

- The need for (certified) standards and reference materials for calibration.
- Traceability to nationally or internationally recognized standards, wherever possible. If this is not possible, documented evidence about the standard's accuracy must be provided.
- Correct labeling with expiration date, storage conditions, date of receipt and initial use.
- Qualification program for standards (tests to verify quality of incoming batches, e.g., identity and concentration, qualification program for suppliers).
- Prepared in accordance with written procedures.
- Prepared from chemicals of known purity and composition.
- Supplier certification to ISO 9001 or the equivalent.

A key question is: When should certified reference materials be used, knowing that the availability of such standards is limited? Regulations and quality/accreditation standards take the limited availability of (certified) and traceable material into account. For example, ISO Guide 25 states:

> Where traceable reference materials are not available or in the case of secondary external standards, the internal reference materials shall be verified as far as technically practicable and economically feasible.

Laboratories have to make the absolute comparison on the best effort possible. This can include the determination of the absolute accuracy by an independent method or by comparisons with other laboratories.

Official Reference Material Programs

Regulatory agencies and quality standard committees have recognized that reliable chemical measurements depend largely on the availability of reliable reference materials with traceability to nationally recognized standards. In response to this demand, programs for the development and supply of material have been initiated in most industrial countries.

United States

In the United States, the EPA monitors and certifies laboratory performance for environmental testing. The EPA provides reference materials free of charge to several thousand environmental testing laboratories through its Environmental Monitoring System Laboratory in Cincinnati (EMSL-CI). While the material is free, the quantities are limited; therefore, it is impossible to use the material on a daily basis [70].

The NIST, formerly the NBS, also offers reference materials through its Office of Standard Reference Material; these can also be used for EPA regulatory programs. Unfortunately, only a limited number of parameters have been certified, and the cost of this material is high. In 1989, the EPA announced the privatization of reference materials. EPA–certified material should be available to anyone. Commercial suppliers offer these materials now as ampuled mixtures, single- or multicomponent mixtures of organics and inorganics suitable for spiking into or onto real samples. The EPA has paid a lot of attention to the quality of commercially available reference materials. The EMSL-CI has entered into various cooperative research and development agreements (CRADA) with several commercial suppliers to produce standards that are EPA approved and labeled as EPA certified [71]. The standards must be produced according to a strict set of specifications, and the purity, identity and analyte must be certified by a laboratory contracted by the EPA.

The EPA feels that in order to provide flexibility within this new framework, it is also necessary to establish a parallel program of third-party certification [72]. This program should ensure the availability of certified reference materials throughout the private sector at competitive prices. In 1989, the U.S. EPA Office of Solid Waste (OSW) requested that the American Association for Laboratory Accreditation (A2LA) initiate a program to certify reference materials. [72] A2LA has worked closely with EMSL-CI and various commercial suppliers of reference materials to develop a series of product specifications. The agreed-on set of specifications is identical to those used by CRADA manufacturers.

Both EPA- and A2LA-certified standards must be verified by a reference laboratory, and the manufacturers of the standards also must maintain a strict quality program similar to or equal to those established by the American Society for Quality (ASQ) and ISO. This includes an extensive internal and third-party audit of their QA program. Registration in the ISO 9000 series quality standards gives U.S. manufacturers the sanction to sell their products to the European marketplace, which relies heavily on these standards.

Europe

The production of certified reference materials also has been a major activity of programs supported by the EU. Previously carried out under the banner of the Bureau Communautaire de Reference (BCR), this work continues in the EU as part of the new Standards Measurement and Testing Programme.

Japan

The development and supply system for reference materials in Japan has been described by Kubota et al. [73]. The system is based on the Japanese Measurement Law of 1991, revised in 1992. Based on this revision, a new traceability system was introduced in 1993. The purpose of the system is to establish and disseminate highly accurate measurement standards that can be linked to sophisticated innovation in both scientific and industrial fields.

The Japanese Chemicals Inspection and Testing Institute (CITI) prepares the primary solutions under the guidance and cooperation of the National Institute of Materials and Chemical Research (NIMC, formerly National Chemical Laboratory for Industry) and the National Institute of Technology and Evaluation (NITE, formerly the International Trade and Industry Inspection Institute).

Traceability to National or Other Well-Characterized Standards

A frequent question with any reference material is: How can I be sure that the concentrations written on the standard's label are correct? This important question has been addressed by many committees when developing standards, regulations and guidelines for quality systems in testing laboratories. Accreditation standards such as EN 45001 and ISO Guide 25 recommend traceability to national standards for tools that are used to calibrate instruments. This also includes chemical standards:

> Reference materials shall where possible, be traceable to international or national standards of measurement, or to national or international certified standard reference materials. Where traceable reference materials are not available or in the case of secondary external standards, the internal reference materials shall be verified as far as technically practicable and economically feasible [67].

Figure 11.1 shows the traceability concept to primary standards that are nationally or internationally recognized standards. Physical standards are traceable to SI units: meter (m)

- **Primary Standards**
  <u>Certified</u> by nationally or internationally recognized institutes (NIST, BCR, NIME)
- **Secondary (certified) Standards**
  Traceable back to primary standards or otherwise verified, for example, through independent test method **with certificate** from manufacturer
- **Working Standards**
  Prepared by user with traceability to primary or secondary standards or otherwise verified, for example, through an independent test method

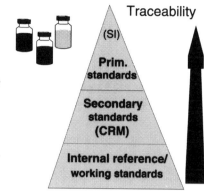

Figure 11.1. Traceabiliy of working standards to secondary and primary standards.

for length, kilogram (kg) for mass, second (s) for time, ampere (A) for electric current, and Kelvin (K) for temperature. In contrast to physical standards in practice, there is no traceability to an SI unit for the Mol (10^{-23}) (at the time of this writing).

For chemical measurement, national primary standards are prepared and certified under the inspection, supervision or technical guidance of national institutes, for example, NIST in the United States or NITE in Japan.

The certified property value or true value, e.g., concentration, together with its accuracy are determined and verified by multiple laboratories using alternative methods. Stability and homogeneity are also determined.

Limited quantities of these primary standards are available, and suppliers of commercially available standards make a direct comparison with their own prepared standards. The comparison must be made in an accredited laboratory using validated reference methods. Suppliers offer these certified standards together with certificates that list the method, the concentrations, the uncertainty and the national standard committee that was responsible for the certified standard.

Laboratories that purchase certified reference standards can use these standards directly for instrument calibration. They also can use them to prepare homemade internal reference standards, which then are used to prepare working standards for day-to-day use.

Requirements for (Certified) Reference Material

Because of the importance of (certified) reference materials in the analytical process for data accuracy, laboratories should ensure that the material meets the following requirements:

- The compounds and concentration should be as similar as possible to the unknown sample.

- The material should be "matrix matched." The matrix of the reference sample should be similar to the matrix of the unknown sample. The same source of error should be encountered when analyzing certified reference materials and unknown samples. For example, if the analysis includes a step where the analyte is extracted from the matrix (e.g., polynuclear aromatic hydrocarbons [PNAs] from a soil sample), the accuracy determination must demonstrate that no or a well-known analyte loss occurs during extraction. If there is one, the amount of loss must be correctly determined for the calibration. Unfortunately, full matrix matching frequently is an unrealistic requirement. There are hundreds of thousands of chemicals being analyzed in all types of matrices, but there are only tens of thousands certified reference materials available.

- The reference material should be homogeneous. Portions of the material will be used from different locations in the container at the same or at different times. Homogeneity ensures that all material at different locations is the same. The difference between sample measurements from different locations should be smaller than the overall uncertainty limits. The material should be checked for homogeneity as part of the verification process.

- If there is a risk of segregation during transport or storage, the material must be rehomogenized before use. Information on rehomogenization should be available from the supplier.

- The certified properties of the reference material and the matrix should be stable. Portions of the material will be used from the container at different times. Stability ensures that all material at different times is the same. The material should be checked for stability as part of the verification process.

- The uncertainty of the value should be estimated for certified reference materials.

The procedure for characterizing the reference material should be validated, the limits should be known and the method should be fully documented and available to the user of the material. Reference materials are certified according to recommendations in ISO/IEC Guide 35 [11].

Preparation and Testing of (Certified) Reference Material

The preparation and certification of reference materials should follow documented procedures in a quality standard environment. ISO Guide 35 [11] gives several technically valid approaches for certifying this material. For a certification, there are essentially three approaches:

1. *Definitive method:* The method must be based on first principles, have very high precision and essentially zero systematic error. An example is the use of isotope dilution mass spectrometry for the characteristics of trace level elements in natural matrix elements. The certification is done in a single laboratory.

2. *Independent measurement method:* Two or more reliable independent methods are used. The method must be proven to give accurate results. The certification is done in a single laboratory.

3. *Interlaboratory consensus method:* A number of laboratories analyze in replicate one or more units of the material being characterized. The participating laboratories may choose their own method or all laboratories may use the same method. The consensus value is usually taken as the mean.

Bolgar [72] and Bunn [74] reported briefly on the manufacturing, testing and QA program as used by AccuStandard and

NSI Environmental Solutions, two suppliers of certified reference materials. Users of certified reference materials should ask their suppliers if they follow a similar process.

1. Raw materials are structurally verified by mass spectrometry, infrared spectroscopy and chromatographic index.

2. Most standards are prepared gravimetrically in concentrations accurate to ±0.5 percent using electronic recording balances calibrated with NIST traceable standards.

3. The analytes contained in a solution are checked for identity, purity and concentration within ±6 percent of the true/reference value by 2 methods against calibration standards from another source, as well as a QC sample from still another independent source.

4. The homogeneity of the production run is analyzed from random samples, 3 each from the beginning, middle and end of run.

5. Stability is established by comparing samples from the lot at 0, 90 and 360 days, followed by tests after 2, 4 and 8 years if indicated.

6. Concentration of each lot is confirmed by an independent accredited reference laboratory, and the results have to fall within ±10 percent of the true/reference value.

Preparation of "Homemade" Reference Standards

Certified standards are quite expensive and frequently unavailable for a laboratory's task. To save costs, it is recommended to prepare reference standards as working standards, also called laboratory standard materials or internal reference materials, in laboratories for daily use, and to calibrate these standards using the certified standards. In this way, the certified standards will last for a long time. Special care must be taken when comparing the working standard with the certified reference material. The method used for the comparison should be validated, and the measurement uncertainty well known. It is a good practice to analyze and characterize the working standard in more than one laboratory, or, if this is not possible, at least within one laboratory by several operators on different instruments over several days to eliminate environmental effects.

Preparing Working Standards from Internal Reference Material or from Certified Reference Material

When preparing a working standard from certified reference material or internal reference material, the following steps are recommended:

1. Develop a procedure for preparing a working standard and follow the procedure.

2. Document details, such as the name of the person who prepared the standard and the date when it was prepared.

3. Label the working standard with the date of expiration.

4. Verify the quality (concentration, identity) with validated methods.

Preparing Internal Reference Material from Certified Reference Material

When preparing internal reference material from certified reference material, the following steps are recommended:

- Develop a procedure for preparing the internal reference material and follow the procedure.

- The composition of the internal reference material should be as close as possible to that of the samples.

- Use pure material; if possible, use a primary standard with traceability to a nationally recognized standard.

- Prepare sufficiently large quantities. The quantity depends on the frequency of use, on the amount used per session and on the stability. For solid materials, 1–10 kg is appropriate if the material is used frequently (1–10 L for liquids). More material should be prepared for reference materials that are used among different laboratories.

- Homogenize solid material.

- Stabilize the material. Most materials change in time due to evaporation or chemical reactions initiated by temperature, light, air or humidity. The values to be certified may, therefore, change. To stabilize the material, it is usually dried either by oven drying or freeze-drying. The stability should be verified with accelerated normal laboratory conditions. Based on such studies, the material should labeled with an expiration date.

- Verify accuracy through comparison with a certified reference material or through comparison with other independent methods.
- Verify the accuracy in a second laboratory.
- Provide information regarding shelf life, storage conditions, applicability, safety precautions and restriction of use.
- Label the working standard with the date of expiration.
- Document the person who prepared the standard and the date when it was prepared.
- Document all details of homogeneity trials, stability trials and the method used for qualification.
- Estimate, document and report an estimate of the uncertainty in the certificate.

Correct Use of Certified Reference Material

The user of certified reference materials should be familiar with all information pertinent to the use of the certified reference material as specified by in-house or external producers. Particularly important are the

- intended purpose in a laboratory,
- period of validity before and after its first use,
- storage conditions,
- instructions for use,
- specifications for validity of the certified properties and
- uncertainty.

Quality Assurance Program

Each laboratory should have a QA program for reference materials and standards, which should be part of the company's or laboratory's quality plan. Steps in this program can include the following procedures:

- Policy on when certified material is required.

- The qualification of the supplier. Certificationtof ISO 9001 or an equivalent standard is strongly recommended; otherwise, a direct audit is recommended.

- Frequency and types of checks of incoming material. Checks can include verification of identity and amounts.

- Registration of the material in a database.

- Handling and storage of the material.

- Preparation of internal reference material and working standards from purchased material.

- Labeling, e.g., expiration date, storage conditions and toxicity.

- Regular checks of the material, e.g., for purity and stability. Reference materials; primary, transfer or working standards; and certified reference materials should be subjected to periodical intermediate checks using a defined procedure.

- Actions to be taken in case the acceptance criteria are not met.

- Incoming tests when the reference material has been prepared and delivered from another laboratory in the same company (this also requires some checks).

- Disposal of used material.

Availability of (Certified) Reference Material

The preparation and QC of a reference material is a complicated process. Liquid solutions are relatively easy to prepare.

Homogeneity is hardly a problem; however, stability testing can be quite extensive.

Solid material is difficult to prepare. It requires extensive blending, homogenizing and analytical measurements, which include complex sample extraction and other derivatization steps. The material must be homogenized, stabilized and checked for homogeneity and stability.

Only highly specialized organizations have the expertise and resources to do the necessary work for this type of material. In the past, government agencies such as NIST in the United States and the BCR in Europe have been the preferred suppliers of reference materials. Because the production of such material can be quite expensive, private companies have tended to stay away from marketing these types of materials [75]. Another reason that prevented private companies from actively entering this market was the competition with governmental institutes. The concern was that they would not have the same level of acceptance.

Today, there are a number of suppliers of laboratory and certified reference materials for the various fields of analyses. Some of them are specialized in a particular field of interest. Suppliers include national institutes such as NIST in the United States and BCR in Europe. They cover several fields and ensure long-term availability due to the large batches of materials produced. The UK Office of Reference Material (ORM) provides a focal point for marketing certified reference materials throughout the world. The ORM is a distributor for the world's leading certified reference material producers, including

- NIST, United States
- EU Standards, Measurements & Testing Programme, Belgium (formerly BCR)
- National Research Council (NCR), Canada
- National Research Center for Certified Reference Materials (China)
- National Institute for Environmental Studies (NIES), Japan

- Laborattoire National d'Essais (LNE), France
- International Atomic Energy Agency (IAEA), Austria

Before purchasing reference materials from a supplier, be certain that the supplier is ISO 9001 certified or certified to an equivalent quality standard.

There are other sources of information on reference material. For example, the ISO Council on Reference Material (REMCO) publishes a directory for reference materials [76]. IUPAC publishes a catalogue of certified reference materials that are available [77]. The Reference Materials Advisory Service (REMAS) of the ORM, which belongs to the LGC in the United Kingdom, provides information on the specification, application and availability of certified reference materials from around the world [77]. Free advice is given to all enquiries on any materials for which data are available, not just those supplied by the ORM. In addition to operating its own database, REMAS also utilizes the information available on the international certified reference material database, Code d'Indexation des Materiaux de Reference (COMAR).

COMAR's database, which is located at the LNE in France, was developed by the collaborative efforts of the LNE, the Bundesanstalt fuer Materialpruefung und -forschung (BAM) and the LGC. It contains data on over 8,000 certified reference materials produced by European, North American and Asian producers. The information stored for each reference material includes the

- name and general description of the material,
- name and address of the producer,
- form of the material,
- certified properties of the material,
- uncertainties of the material,
- date of certification and
- fields of application.

The COMAR database can be purchased from the ORM [78].

Summary Recommendations

1. Develop a policy and procedures for (certified) reference materials. (When is which quality of material required? Which type of traceability is required?)

2. Develop procedure for qualification of (certified) reference materials and suppliers.

3. Develop procedures for preparing homemade reference and working standards.

4. Purchase certified reference materials (if available), and prepare relatively large amounts of in-house reference and working standards.

5. Develop a QA program for (certified) reference materials.

12. People

What will be discussed in this chapter?

1. The key issues when recruiting new people
2. How to qualify people for their jobs
3. The training methods available
4. How to document evidence of qualification

The single most influential factor in acquiring accurate and reliable data is the hiring, training and managing of qualified people. Regardless of all the documentation and automation available in a laboratory, if people are not properly qualified and motivated to handle all laboratory activities, one will not obtain consistently good analytical data. For example, the best computerized systems cannot generate accurate and reliable data if the operator makes wrong entries because he or she did not receive sufficient, job-oriented training. In order to perform the job well, each employee must have a background combining education, experience and training.

Consequently, there is a chapter about people in all regulations and guidelines and quality and accreditation standards. The passages are usually brief and contain sentences such as the following in the OECD GLP Principles [47]:

> A test facility has appropriately qualified and experienced personnel and there are documented training programmes including both on-the-job training and, where appropriate, attendance at external training courses. Records of all such training should be maintained.

Recruiting Qualified People

It is well beyond the scope of this book to teach managers how and where to recruit people. Most organizations have reasonably well-structured qualification requirements and well-established hiring processes. However, because this is such an important topic, readers should at least consider the following four recommendations:

1. When recruiting new people for a specific job, it is very important that the person has the right knowledge and technical qualification for the specified job, as well as the right personality to fit into the laboratory's environment. The technical skills may have been obtained through education, and/or experience in a specific job. These are well-known criteria usually used for hiring new people. However, in today's rapidly changing world with changing working tools, it is equally important that the person have proven flexibility to learn new techniques, processes and tools and to take over new tasks. For example, being willing and able to work with modern on-line communication media like the Internet and the Intranet is very important, even for those jobs that currently do not require such knowledge.

2. Teamwork is becoming more and more important. With globalization, tasks are shared not only within a single laboratory but within a company across divisions, countries and continents. Frequently, part of the work is outsourced to other companies, and processes are improved through working closely together with suppliers and customers, all of which requires excellent communication. Therefore, evidence of good teamwork and communication skills is of utmost importance.

3. Hire people who love to do the work they are supposed to do. Make sure they are enthusiastic about working in an analytical laboratory. This natural motivation can be assessed during interviews and when the candidate(s) is giving a presentation on his or her previous work.

4. One should always consider the subsequent step in respect of the candidate's next possible job or career plan. It is very unlikely that somebody will remain in the same job forever. If the candidate may have to take over

supervisory responsibility, relevant criteria should also be applied when interviewing for a job that does not have supervisory responsibility.

The typical process for hiring new people is as follows:

1. Description of the job, including its "must have" and "want" requirements
2. Job posting
3. Screening of applications
4. First round interview
5. Second round interview
6. Generating and signing the contract

Searching for Candidates

Finding ideal candidates can be very difficult for certain specialized jobs. All types of sources should be used, such as the following:

- Internal job boards
- Posting in company internal newspapers and the Intranet
- Posting in external newspapers, magazines and the Internet (electronic bulletins)
- Posting at schools, universities and institutes
- Using state employment services (unemployment office)
- Using private employment agencies (search companies). These can be very effective and may work well when looking for people with specialized skills. The employer describes in great detail the job position and "must have" requirements. The agencies contact candidates and ask them if they are interested in the job. Usually, the search company charges a fee of up to 30 percent of the annual salary if the hiring is successful.
- Advice from company internal and external colleagues
- Posting at symposia and exhibitions

Posting at exhibitions and conferences can be quite efficient. For example, the PITTCON conference, probably the world's largest conference and exhibition for analytical instrumentation, has an employment bureau where employers can post jobs using employer registration forms and where candidates can apply for jobs. This service is free of charge. Fax forms for employers and candidates can be requested by phone and should be sent to the employment bureau about four weeks prior to the conference. Each candidate and each supplier receives a registration number. The files of all employers are available for inspection in the candidates' room, and the files of the candidates are available in the employers' room. Both rooms have a number of tables, and each table contains a complete series of binders with all of the candidates' or employers' forms. If an employer is interested in a candidate, a message can be submitted to a message box using the assigned registration numbers as identifiers. Similarly, candidates can show their interest in a posted position by submitting the message card to the employers' message box. If an employer wants a complete resume from a candidate, a sticker is placed on the message card. If an employer wants to interview a candidate during the conference, an interview scheduling form is filled out and the employer receives confirmation of the interview date and time. This free-of-charge advertising can be very efficient, because the jobs are announced to many candidates, and several interviews can be set up within a few days.

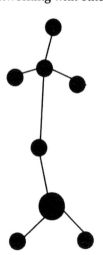

Most people get their jobs using informal methods through networking with others

Most people get their jobs using informal methods. This may be through networking with people you know and by contacting the candidate directly, or through his or her supervisor, if the person is from your own company. If you have to hire people more frequently, develop and maintain a list of contacts whom you may ask if a needs arises. Also keep a list of candidates whom you should regularly contact and talk to about their interest in working for your department. In this way, you can watch their ongoing performance; when you really have a job opening, the hiring process is then very rapid. You should use all types of resources to develop such lists: friends, colleagues from your own company and from outside your company, scientists you may meet at conferences or symposia, and even competitors whom you may meet at exhibitions.

Job Posting

Any job advertisements should have an attractive format and contain a clear indication on how to obtain further information and who to contact to make an appointment for an interview.

In general, any job posting should include the following:

- Description and location of your organization
- Title or summary description of the position
- Description of tasks and responsibilities
- "Must have" and "want" requirements
- Contact person, phone number and address

Screening of Applications

The first step in selecting possible candidates is the screening of written applications. It is recommended to use checklists with check items for the following:

- Is the documentation complete?
- Can any conclusions be drawn from the letter regarding personal style and how he or she may fit into the lab environment?
- Does the candidate meet the "must have" requirements?
- Which "want" requirements are met?
- Is there any indication of criteria such as work attitude, creativity and communication capability?

The Interview

The interview is the most important step in the process and should be discussed in a little more detail. Some of the questions that should be answered before an interview takes place are as follows:

1. Who should participate in the interview(s)?
2. What questions should be asked at the interview(s)?
3. What conclusions can be drawn from the answers?

Recommended interview participants:

- First line supervisor
- At least one manager level above
- One or two colleagues of the supervisor

- One or more colleagues of the prospective new employee
- Member of personnel department

Including future colleagues of the candidate is important to ensure acceptance and a smooth integration in the laboratory. The number of interview candidates should be selected in such a way that the interview process does not last for more than one day. Before each interview starts, every interview participant should be well informed of the job description, the "must have" and "want" requirements and the candidate's background. Interviewers should get a good idea of the candidate's technical background as well as on some personal characteristics. For example, from the average time spent at one particular company and in one specific job, one can draw conclusions about the willingness to work at one company for a longer period of time or the candidate's flexibility to take over new responsibilities. The lead interviewer, usually either the direct supervisor or the representative of the personnel department, should talk individually with each interviewer and provide some recommendations on what each person should cover as a minimum. This will avoid redundancy of questions and increase the amount of information interviewers receive. However, on certain critical issues, it may be valuable for the same questions to be put to the applicant by different interviewers.

Every interview situation is so different that it is difficult to give general recommendations. Here are just a few considerations that an interviewer may wish to bear in mind. What one really wants to find out is how the candidate will perform in the future. As the future is very difficult to predict, the most accurate information can be obtained from the candidate's behavior in similar situations in the past. Therefore, most questions should be related to how the candidate worked to solve specific problems in the past. For example, if you are looking for a creative person, ask the candidates to provide relevant examples from their work in previous positions. Similar questions should be asked on flexibility, self-motivation, problem-solving skills, work ethic, broad interests, supervision (if this is a "must have" or "want" requirement) and broad personal skills. If a candidate claims to have specific skills, always ask for examples or references.

The overall key success factor of an interview is to ask specific questions and get candidates to talk about their skills, experiences, work habits and professional attitudes. A good rule of thumb is that during the interview you should talk for about 20 percent of the time and the candidate should talk for about 80 percent. Borchardt [79] introduced this 80/20 rule to the job interview practice. A good way to encourage the applicant to speak is to ask open-ended questions beginning with "why," "how" and sometimes "what." These cannot be easily responded to with a simple "yes" or "no." There may be follow-up questions to specific important topics, such as, "Tell me more about . . . ."

Besides the questions regarding technical qualifications, answers should be obtained to questions relating to personal skills such as

- flexibility,
- ability to learn,
- ability to take risks,
- ability to deal with problems,
- ability to work in a team,
- ability to listen,
- creativity,
- ability to meet critical time schedules,
- verbal and written communication skills and
- work attitude.

Typical questions for the candidate are as follows:

- Why did you apply to our company?
- What do you know about our company?
- Why did you apply for this job? What is attractive and what is not?
- What are your qualifications for this job? Give examples to prove your statements.

- What do you like or not like about your existing job?
- What work style do you prefer? Would you prefer to work independently or with clear guidelines?
- Would you rather work with others or alone?
- What would be your ideal work group?
- Have you ever had to motivate coworkers? Describe how you did this.
- What form of communication do you prefer and why?
- Have you worked on major projects in any of your current or previous jobs? What were your contributions to the project?
- How many projects can you handle at one time? Give examples.
- What is your added value to the new job? What can you do that others cannot? Give examples for your statements.
- What are your strong points? What are your weaknesses?
- What is your learning style (self-study, classroom trainings)?
- How do you keep informed professionally?
- What important trends do you see in our industry?
- What do you think about your current/former supervisor?
- What is your definition of success? How would you describe success?
- How do you make decisions? Describe the process you go through to make decisions.
- What is your motivation to change your job?
- Why should we hire you?
- Do you have any examples that prove your creativity?
- Describe a situation from your current/previous job where you experienced and solved a serious problem.

- What are things you find easy to do?
- What are the things you find difficult to do?
- What kinds of decisions are most difficult for you?
- Have you ever had to work under pressure and deal with deadlines?
- How do you deal with unexpected events in your job?
- Under which circumstances would you work overtime and on weekends?
- How long would it take for you to make a contribution to our company?
- What do you think are the most important success factors for the new job?
- How long do you plan to stay in the new job, and what is your long-term goal (5 years)?
- How would you rate your writing skills as opposed to your oral communication skills?
- What kinds of people do you like to work with?
- Do you prefer delegation or hands-on control? (The answer to this question will help to determine if the candidate will fit into your work environment.)
- What type of supervising would you prefer (detailed, independence)?
- You have been in your previous position an unusually long period of time—why is this so?
- You have changed jobs more frequently than usual—why is this so?
- What are your outside interests? Do you have any hobbies?

Answers to these questions can best be obtained by talking about current and previous work. The candidate should always be asked to give examples for any statement.

If an important task of the applicant will be to give presentations, the candidate should be asked to give a presentation in front of an audience. The topic of the presentation should be selected by the applicant, and the presentation should not last for more than 15 to 20 minutes, including questions and answers. The presentation style, the logic behind the presentation, the level of excitement, the way difficult questions are handled and, last but not least, the ability to meet the time schedule are all important criteria for judging technical, personal and oral communication skills. Such a presentation also has the advantage that more people than just the interview team can listen to the presentation and give inputs on the applicant's personality and qualification. In this way, future colleagues who will be in direct contact with the new employee can be included in the hiring process so that they too can feel good about the candidate.

After the interviews, all interviewing team members should meet to discuss the outcome. Everybody should give their overall impression and findings from the interview before the lead interviewer opens a discussion on the final selection.

Defining and Communicating Job Descriptions, Tasks, Responsibilities and Desired Outcome

Make sure that the job is accurately described and that tasks and responsibilities are well understood and accepted by the chosen person. This holds for both new and senior people. People may become very discouraged if they do their utmost to do an excellent job and later discover they did the wrong thing. A good job description with clear written expectations on the goals and standards is of utmost importance. According to ISO Guide 25 [67], the responsibilities of staff in calibration and testing laboratories include

- performing tests and calibrations;
- planning of tests and calibrations and evaluation of results;
- development, validation and modification of methods;
- professional judgment; and
- managerial duties.

Besides the business tasks that are usually included in the laboratory's objectives, each person should also propose and discuss with the supervisor some personal development goals for a given time frame. Because responsibilities and job tasks can change over time, these should be reviewed on a formal basis at specific time intervals.

Monitoring Progress and Providing Feedback

The person's development process in the job and the extent to which the previously specified objectives have been met should be monitored. Instant and regular feedback should be given to employees on how they are doing and how they could improve their performance. Usually, more formal performance reviews are done each year (Table 12.1). Evaluation items include quantity and quality of work, communication and teamwork, creativity, customer satisfaction and work safety. Other items to be discussed include the employee's personal long-term goals and possible barriers preventing good performance. Many companies supply forms to be used for preparation and during the meeting itself and to document the results and objectives for the following year. It is important that there is consistency between the person's objectives and supervision on the employee's appraisal. If this cannot be achieved, other people could be invited to a future meeting (e.g., the next level manager or a member of the company's workers' council).

Regular feedback and performance evaluation meetings encourage a more positive attitude about the employee's job. The supervisor develops a better understanding of the employee's strengths and weaknesses and about training needs and career plans.

Continuous good work should be rewarded. Promotions or financial compensation are ways of doing this, but analysts may have other wishes: prestige among colleagues, visibility in a larger organization and chances to learn new things. Frequently, people also want to share their success with their families.

Training

Well-trained personnel are among a company's most valuable assets. Proper training not only builds skills as required for the job but also builds confidence. Training should go beyond

Table 12.1. Form for joint annual review	
Name	
Job Title	
Department	
Date	
Main job duties and responsibilities (refer to job description and goals/measures from last year)	
Main achievements	
What did go well; what did not go well; why was this?	
What was different from previous years?	
Working relationships and performance	
Points that may be discussed Quality of work Quantity of work Communication Work attitude Motivation Safety	
Summary of action plan	
What should be done differently in the future?	
Development goals, short term and long term	
Main business goals/measures for next year	

analytical instrumentation and methods. It should also include safety and personal skills, such as improving creativity, communication and teamwork. As a rule, training should account for 15 to 20 percent of an analyst's time. Initial and ongoing training should be given on

- analysis techniques;
- equipment;
- methods and procedures;
- regulations and quality standards;
- environmental, health and safety;
- teamwork;
- improving communication; and
- improving creativity.

Unfortunately, training is traditionally the last item addressed by supervisors and the first cut when time or budgets run out.

A high-quality people training system should be in place to ensure that

1. laboratory staff have sufficient entry training and permanent ongoing training to keep up-to-date with constantly changing instrument capabilities and regulatory and quality standard requirements and

2. all education and training activities are documented.

A problem may occur with training if personnel are brought in on a short-term basis from another department. Such people may have adequate experience and knowledge for their permanent job but not for the one actually performed. It is important that there is documented evidence that the current job can be performed with sufficient quality. This is relatively easy to do if the job is similar. A statement about the similarity, together with a reference to the qualification documents in the other department, is sufficient. If the job is different, there should be full qualification documentation based on trainings for the new job.

The ASTM *Standard Guide for Training Users of Computerized Systems*

> ASTM E 625 - 87
> (Reapproved 1992)
>
> Standard Guide for Training Users of Computerized Systems
>
> 1992
>
> Reference 98

The ASTM has developed a standard guide [98] for training users of computerized systems that can be used as a model for trainings on other topics. The guide presents a six-phase plan:

1. *Defining the training program:* Develop a brief statement of what is to be accomplished and why.

2. *Setting functional requirements:* This phase identifies and defines all inputs and outputs. It includes definition of the target audience by position, experience, education and skill requirements. Users of the training should be heavily involved in this phase.

3. *Producing functional design:* Using the functional requirements document as a basis, produce detailed descriptions with flow of information and timing diagrams. Identify information needed and define where trainers can obtain the information. Include training elements to provide practical experience in applying the new knowledge. Provide some means of testing whether the essential information and required skills have been communicated and understood.

4. *Implementing design:* Select components and develop implementation, operation and maintenance procedures. Select training methods on the basis of conformance to the functional design and costs.

5. *Implementing the training program:* Order materials, reserve positions in outside trainings and make travel arrangements. Prepare lecture notes and hand out material. Prepare course examinations for evaluating the training program.

6. *Evaluating the training:* Compare what is accomplished against functional requirements to determine how well the requirements are met. Inputs will be used to improve future training sessions. Evaluations can be made based on interviews and by completing evaluation forms. If possible, use the same evaluation form over a longer period of time and for different kinds of trainings. This makes it easy to monitor long-term success and to make comparisons between different trainings.

Methods of Training

There are different training methods for different circumstances and requirements. The methods, tools and possible trainers are summarized in Table 12.2, and the methods are discussed in more detail below.

Classroom Training

Classroom training occurs in traditional seminars that are offered by instrument vendors, consulting companies and technical organizations. The advantage is that a large number of people can be reached at one time, and direct feedback comes from the attendees in the form of questions that can be answered directly by the instructor. In addition, students can share their experience with others. The disadvantage is that such training courses may not be scheduled when required.

Individual and Small Tutorial Group Training

Small group trainings are offered by instrument vendors and consulting companies. They can be tailored to the knowledge

Table 12.2 Training methods, training tools and training organizations

Training methods	<ul><li>On-the-job</li><li>Individual instruction</li><li>Classroom training</li><li>Self-study</li></ul>
Training tools	<ul><li>Paper</li><li>Audio</li><li>Video</li><li>Slides</li><li>Computer-based</li><li>Multimedia</li><li>Intranet and Internet</li></ul>
Who delivers trainings	<ul><li>In-house specialists</li><li>Instrument vendors</li><li>Scientific organizations</li><li>Private organizations (consultants)</li><li>Schools and universities</li></ul>

and special requirements of the student. They can be scheduled at a time when the instruction is needed. The students' questions can be answered immediately by the instructor. The disadvantage is that they are usually quite expensive.

Computer-Based Trainings and On-line Tutorials

Tutorials or computer-based trainings are programs that are part of the software and are relatively inexpensive. They are tailored to the specific training needs for the software. Interactive operation makes them highly efficient. They can be used as the operator progresses on the software. The disadvantage is that there is no direct feedback to the instructors or immediate answers from the instructors if there are any questions.

Videotapes

Videotapes are supplied with or are otherwise available for some software packages. The advantage over slides/audio is that moving pictures can better illustrate complex technical processes. The disadvantage is that video recorders are not usually readily available in offices and not enough videos are available to demonstrate the tasks for which they would be most useful, e.g., complex technical processes.

Multimedia CD-ROM

CD-ROMs are the most modern training tools. Text, audio and video pictures make them ideal to convey all kinds of information: analysis techniques, complex technical processes, as well as instrument operation. The disadvantage is that they are expensive to produce and not many CD-ROMs are available for equipment training (as of 1998).

Standard Operating Procedures

SOPs can be a useful training tool for operating and maintaining an instrument. Typical step-by-step instructions make it easy to learn the instrument functions. There are usually no additional costs for the acquisition or development of training material because the SOPs should already exist. The disadvantage is that they provide no opportunity for personal interaction.

Textbooks

Textbooks are traditional, self-paced training tools. They are relatively inexpensive and are most convenient because they are easily transported and readily available when and where they are needed. They are useful for obtaining information on regulatory and quality standard compliance. They are also a useful source for reference material and checklists. The disadvantage is that they provide no opportunity for personal interaction, and complex processes are difficult to communicate.

On-the-Job Training

Learning by doing sometimes appears to be least costly, but it also requires good supervision. Execution should be planned in as much detail as classroom trainings. On-the-job training is not appropriate for new technologies.

On-line Training

On-line tools like the Intranet and the Internet may become an other important training tool in the future, when audio and video communications will be more practical.

The Ideal Training Tool

The ideal training tool depends on the type of training task, on the availability of different trainings in the user's geographic area and on the urgency of the training. A combination of different trainings, for example, with books or videos as prestudy material, followed by classroom trainings, is most practical.

Documentation of Trainings

Each training activity should be documented with content, dates and location. An example for a template is shown in Figure 12.1. Documented training activities should include on-the-job training; individual instructions on specific tasks; official classroom trainings; short courses by scientific organizations; training on special techniques and instruments by instrument vendors; and self-study training by reading books, video trainings or multimedia trainings.

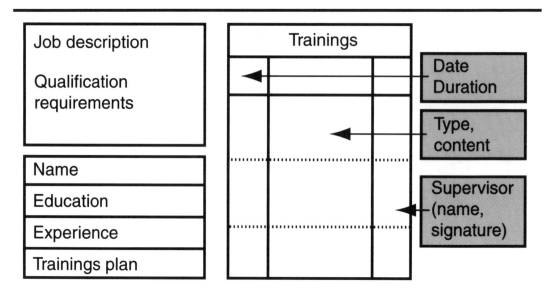

Figure 12.1. Form for job descriptions and training records.

Frequently asked questions are as follows:

- How should the success of a training be evaluated and documented?
- Where can I obtain certification of the quality of the training organization and the instructor? Who finally qualifies the trainer? What documented qualification should a training organization have?
- Which training activities regarding length should be documented: hours, half day, full day, a week?
- Is self-study appropriate? How should it be documented?

For all trainings, certificates on the successful participation should be available and signed by the instructor. Training certificates are often included in distributed material and provided at the beginning of the training. This is not appropriate and does not prove that the person participated in the full training. Certificates should be distributed at the end of the training, preferably after the attendees have successfully passed an exit test.

The training organization and the instructor should be able to demonstrate evidence of qualification. This may be via training records or other proof of competence, for example, a number of papers published or a number of presentations given on the topic where the speaker has been invited. The training organization should be ISO 9003 certified or should have documented evidence that follows another quality standard.

Some people learn best by reading books, watching videos, working through on-line tutorials or using modern multimedia tools. The question is on how to document the success of these trainings. The author's recommendation for all self-study activities would be that the person's supervisor should sign off on the successful completion of these trainings. Supervisors are responsible for the qualification of their people. They should know the training needs and should be fully aware that successful trainings may it be received in official training courses or through individual self-studies.

Regarding the documentation of trainings of different lengths, the author recommends documenting all training activities that last for one or more days.

Besides documentation for training, it is recommended that for each job within a company, a file is kept with a clear job description and information on education and training requirements. It is also recommended that the company maintain an official personnel file for each employee that holds information on education, experience, ongoing training activities and participation in proficiency testing programs. This file should be updated as necessary. Because this documentation should be available for internal and external inspectors, it is recommended to keep information on the job description, job requirements, skills, education and training separate from other, more personal and confidential information, such as results of performance evaluations. Laws on privacy also often require that access to any personal file be restricted.

Summary Recommendations

1. Allocate plenty of time for recruiting new people.
2. Develop an ongoing network of candidates among people you know.
3. Develop a policy and procedures for identifying training needs and providing training of personnel.
4. Describe job function and responsibility for each person.
5. Describe the person's education and experience related to the job function and responsibility.
6. Describe training requirements (gap between job requirements and current education or experience).
7. Develop a short- and long-term training plan for each person.
8. Document any training of personnel.
9. Review training needs every year.
10. Evaluate the success of the training.
11. Keep records of relevant competence, education and experience of all personnel concerned with instrument qualification, sampling, measurement and data evaluation and reporting.

13. Proficiency Testing for External Laboratory Qualification

What will be discussed in this chapter?

1. Operation of proficiency testing schemes
2. Advantages for laboratories
3. Evaluation procedures
4. Who should participate in proficiency testing
5. Frequency of testing
6. Limitations of proficiency testing

> R.E. Lawn
> M. Thompson
> R.F. Walker
>
> *Proficiency Testing in Analytical Chemistry*
>
> 1997
>
> Reference 80

Proficiency testing by interlaboratory comparisons can serve two purposes:

1. Test an analytical method for effectiveness and ruggedness done on an irregular basis.
2. Assess the regular technical competence of participating laboratories to generate comparable analytical data.

This chapter will focus on the second item. Proficiency testing can enable a laboratory to compare its performance with that of other similar organizations and provide independent evidence of the validity and comparability of its data, i.e., to qualify the laboratory for specific analysis. Proficiency testing is complementary to the analysis of in-house QC samples. While QC analyses serve as a tool for internal QC, proficiency testing is a tool for external QC. Successful participation in proficiency testing schemes proves that the entire analytical quality process is working well. This holds true for the analytical method, the equipment hardware and software, reference materials and people. Bodies assessing the technical competence of testing laboratories, such as accreditation and certification bodies, use the results of proficiency testing in their assessment.

ISO/IEC Guide 43 [81], as developed by ISO/CERTICO in response to a request arising from the International Laboratory Accreditation Conference (ILAC82), covers guidance on the development and operation of proficiency testing. At ILAC94, a revision of the guide was presented in draft form entitled *Proficiency Testing by Interlaboratory Comparisons* and includes statistical guidance on the treatment of data received from proficiency testing [82].

Procedure

In a typical proficiency testing scheme (Figure 13.1), portions of a well-characterized test material are distributed by the organizer, on a regular basis, to participating laboratories for analysis. The laboratories analyze the samples using methods and standards usually applied for that sample and send the

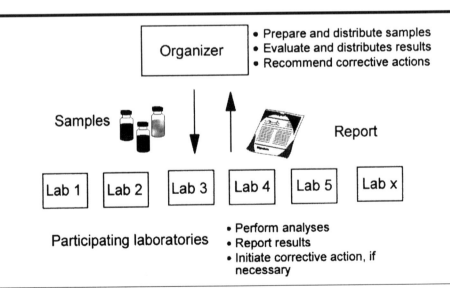

Figure 13.1. Typical proficiency testing program.

results back to the organization that distributed the test material. Each laboratory's result is then compared to the true value for the test material concerned. Depending on the degree of agreement to the true value, the laboratories are scored and receive a report that enables them to review how well they have performed in the test. The results are confidential to the laboratory and the organizer, but clients of the laboratory and the accreditation body may request the test results.

Typically, calibration standards are not sent with the sample, and the analytical methods are not mandated. However, laboratories are advised to report the method because this may be used to obtain information if the method itself is a source of the deviation to the true results.

Evaluation of Proficiency Testing

The proficiency testing process should follow a protocol that has been developed by a collaboration of scientists from many countries under the joint organization of ISO, IUPAC and AOAC International [83]. One of the goals of the procedure is to find a way to convert the data of the laboratories into scores that are easy to understand and of universal applicability. The method recommended in the protocol, therefore, is based on simple statistics with no scaling. Each result (x) is converted into a "Z" score according to the equation:

$$Z = (x - y)/\sigma$$

where y is the assigned value, the best estimate of the true concentration of the analyte. Sigma (σ) is the target value for the standard deviation of values of x. It describes the previously specified acceptable variability between laboratories and is related to the ruggedness of the analysis method. Z scores between ±2σ will occur in 95 percent of all cases and are regarded as satisfactory. Z scores between 2 and 3 are considered to be questionable and will occur in 5 percent of all cases, but those outside the range of ±3σ are considered unsatisfactory. Results are plotted to visualize them as easily as possible and are sent to each laboratory. An example is shown in Figure 13.2.

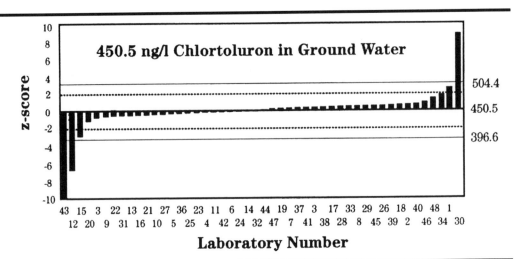

Figure 13.2. Graphical plot of proficiency testing results: 44 laboratories are satisfactory, 3 are unacceptable.

Who Should Participate in Proficiency Testing?

Proficiency testing is most appropriate for laboratories performing routine analysis of sample types that are also analyzed in other laboratories. Examples are environmental, clinical and food testing laboratories where application specific proficiency testing schemes are frequently available. Proficiency testing is required by accreditation standards if proficiency schemes are available. Accreditation bodies encourage testing laboratories to participate in these programs as part of their quality system. A laboratory should look for organizers that use test samples that are the same or similar to typical samples analyzed in the laboratory.

Frequency of Tests

An important parameter in proficiency testing is the frequency of testing. If done too frequently, costs for distributing the samples, performing the tests and collecting and evaluating the data are too high. If done too seldom, a laboratory may not obtain feedback fast enough if there is a problem with analysis. One important criterion is to distribute the proficiency test results early enough prior to the next testing so that laboratories can react to the previous tests if there have been any problems. Once the system is established, one to three tests per year should be sufficient.

Testing Material

A key element in the success of any proficiency testing program is the quality of the sample. This means the laboratory must obtain a sample with the true concentration at the time of the analysis. The variability between test samples delivered to laboratories must be less than the expected standard deviation for the test. This requires three steps:

1. The true concentration must be determined for the entire batch.

2. The batch must be divided into representative parts.

3. The sample must be stable under the conditions stored and shipped.

Some schemes use the mean value obtained from the participants to assign the true or consensus value. Such schemes will make laboratory results consistent, but not necessarily true. Ideally, assigned values should be defined by specialist laboratories using well-validated methods and material that can be traced back to national or international standards. The uncertainty of measurement should be minimized and under control.

Advantages for Laboratories

Participating in proficiency testing schemes can be quite expensive; therefore, laboratories should make a thorough judgment before deciding to participate. Advantages for a laboratory are as follows:

- External and independent assessment of data quality for specific tests.

- A means of demonstrating the data quality to customers, accreditation bodies and regulatory agencies.

- A motivation to improve analytical quality.

- Information on the performance characteristics of analytical methods and the quality of reference materials.

- A laboratory experiencing difficulty with a particular analysis can often seek advice from the scheme organizer to improve its processes. Investigations can result in improved methods being introduced and, therefore, produce more accurate data.

- Compliance with accreditation standards.

Performance Improvements

The main question is whether participation in proficiency testing will improve a laboratory's performance. Patey [84] reported a dramatic rise in the overall performance of laboratories after participating in proficiency tests. This observation was based on over 12,000 sets of results with 183 laboratories from more than 30 countries participating in the tests over 4 years. While at the outset only 60 percent of the laboratories sent in satisfactory data, this value has increased to 90 percent.

Remaining Issues

Aside from all of the benefits for the laboratories, there are still some issues regarding proficiency testing:

- The quality of the results may not be representative for the laboratory. In order to look good, the laboratory may treat the proficiency sample differently from normal routine work. For example, multiple analyses may be done and average results reported, or only the most experienced operators and best instruments may be selected. Therefore, there is some doubt that the reported results are really useful to judge a laboratory's competence for the specific routine analysis.

- The proficiency sample is usually a real sample, not just a spiked standard. Because of a lack of knowledge as to the "true concentration" of the analytes, this may be determined by a consensus of laboratories that have determined the concentration. There is always a risk that this assigned result may not be correct. In this case, a high score will always make the laboratory results consistent but not necessarily true.

- The proficiency sample may not reflect the average concentration as distributed to other laboratories if errors have been made when dividing the sample.

- There are not enough organizers.

- Frequently, it is difficult to find the right sample that corresponds to a laboratory's competence.

- A good performance with a specific type of analysis does not necessarily indicate a good performance on other analyses. It is not always possible to distribute test materials that exactly resemble a laboratory's routine test samples.

- Proficiency testing schemes are relatively expensive, in terms of both organizational costs and the time spent by participating laboratories. Especially for small laboratories with a low sample throughput for a particular analysis, the costs are relatively high compared to the revenue from that specific analysis.

- Most proficiency tests are done on a more local basis with less international focus and with a relatively small number of laboratories. The real value of a specific scheme would be to demonstrate international comparability.

Summary Recommendations

1. Evaluate the need for proficiency testing.
2. Select the right proficiency testing scheme.
3. Decide on frequency of tests (consult accreditation body and organizer).
4. Evaluate your performance in comparison to others.
5. Discuss results and possible improvements with the organizer.

14. Audits

What will be discussed in this chapter?

1. Objectives of audits
2. Mistakes others made
3. Organization of audits
4. Differences between horizontal and vertical audits
5. Aspects to consider in preparation for an audit
6. How to conduct an audit
7. What should be included in an audit report

Audits are a key element of any quality system. Their objective is to evaluate activities and existing documentation to check whether these meet predetermined internal and/or external standards and/or regulations or customer requirements. There are several types of audits:

1. *Internal audits* are conducted on a regular basis to check whether particular departments and individuals adhere to company policies, standards and procedures. These are a requirement of most regulations and accreditation standards relating to analytical laboratories.

2. In *second-party audits,* a purchasing company audits the supplier. These are commonly used to check whether a supplier meets the purchaser's requirements.

3. *Third-party laboratory audits* are used to ascertain whether a company or laboratory complies with national or international quality standards, such as the ISO 9000 and EN 45000 series, or to check whether the company is competent enough to perform analyses, as specified in

---

ISO 10011-1

Guidelines for Auditing Quality Systems

1990

Reference 4

contracts with clients. Regulatory agencies inspect laboratories to confirm their compliance with GLP, GCP and GMP regulations.

Besides checking compliance with internal and external standards, there is a second and even more important aspect of internal and external audits: They can be used to help improve processes and to establish a better system for the benefit of laboratory owners, employees and customers. If the procedure is done correctly, laboratory departments can learn extensively from auditors and inspectors because, as outsiders, they may contribute useful expertise and tips on how to improve certain quality aspects. Laboratories may also benefit from the mistakes made by other laboratories. Sometimes, inspectors publish their observations on problems and deficiencies noted during inspections of pharmaceutical analytical laboratories [85–88]. Occasionally, U.S. FDA inspectors also give a summary of their findings at validation seminars [89, 90]. FDA inspectors record observations on special forms, which are, to some extent, published on the Internet [91]. Before various audit techniques and audit items are discussed, some observations and findings taken from laboratory audits will be examined.

Observations Reported During Inspections and Audits

Observations, as published by inspectors, can be a useful source of preparatory information for laboratories. They may also help to improve a laboratory's work, as they enable the laboratory to take note of and avoid the same mistakes and, therefore, to implement their own processes in a better way. Table 14.1 includes a summary of audit/inspection findings. The observations have been derived from a variety of sources:

- publications by inspectors [85–88],
- presentations given by inspectors [89–90],
- FDA 483 observations and comments [91] and
- the author's own experience.

Table 14.1. Problems found during laboratory audits

Equipment and Software Validation and Testing

- Qualification and validation arrangements for the system were poor with a lack of formal protocols, acceptance criteria, testing procedures, records, reviews, error handling arrangements, formal reporting and signing-off.
- The firm had no formal written validation program covering inspections, checks and testing to demonstrate proper performance.
- Validation test logbooks gave an incomplete record without explanation.
- Retrospective validation/evaluation requirements for poorly documented existing systems were not defined or targeted for completion.
- There were no data to ensure that the HPLC computer system/software had been validated.

Equipment Calibration

- There were no written procedures containing specific directions and/or limits for accuracy concerning the calibration of laboratory instruments and equipment used in finished products testing, including HPLC instrumentation, UV/visible spectrophotometer, AA spectrophotometer, ovens, and so on.
- Written procedures did not ensure that the dissolution apparatus was calibrated each time the shaft was changed for paddles and baskets. The SOP required calibration of the unit only once per year.

Internal audits

- There were no internal routine system audits by QA.
- Noncompliance reports for process or product parameters were lost in a mass of routine acceptable data in piles of printouts.

Reporting of Data

- The final report was undated and unsigned and showed erroneous values.

Method Validation/System Suitability Testing

- There were no written procedures for the validation of analytical procedures and test methods. The suitability of such testing methods was not verified to ensure that they were compatible with conditions that exist in this facility (equipment, environment, personnel, etc.). Written procedures did not specify in-house limits for variable operating parameters that could affect accuracy, reliability and reproducibility of test methods adopted from standard references and compendia.

Continued on next page.

Continued from previous page.

- HPLC analytical method validation of the two methods reviewed found the following: no determination of precision, specificity and accuracy, with no statistical analysis on the nonlinearity of data.

Analytical Quality Control

- Management had not established time limits or limitations for the number of HPLC samples/injections that can be run consecutively without reinjecting reference standards.
- System suitability standards and reference standards were not reinjected at the end of lengthy multiple-sample runs to ensure that the integrity and reliability of the procedure was maintained.

Audit Trail

- PC–controlled instruments with general purpose operating systems and editing screens enabling changes to be made to analytical procedures and data were not always well controlled.
- Laboratory information management systems did not have forced logging of original method and data (before and after changes) with reasons for change, authorization, identities, times and dates.

Personnel

- There were no detailed user-training records.

Vendor–User Relationship for Purchased Computerized Systems

- There was an absence of formal contracts covering technical, QA and GMP requirements for suppliers providing software and systems and specifying limits of responsibility.

Planning and Implementation of Internal Audits

Needless to say, not all laboratories can be audited for all items at once. Over a certain period of time, however, all items should be checked in all laboratories. Therefore, audits should be conducted according to a long-term plan. The objective is that all departments or laboratories be audited for all items over the planned period. Priorities of the audits can also be set based on current trends and regulatory focus.

There are two ways to achieve comprehensive coverage—the horizontal and the vertical approach.

Horizontal Audits

Using the horizontal approach (Table 14.2), all departments are audited, in detail, for the same item at one particular time; for example, for organization and methods or equipment. In a subsequent audit, other items are checked. Horizontal audits may reveal only some of the weaknesses that may exist in the quality system.

Vertical Audits

In a vertical audit, all or a selected number of different items are checked at one particular time. In practice, not all laboratories are audited at the same time; they are audited

Table 14.2. Horizontal audit schedule

Horizontal (all labs)	99/Q1	99/Q2	99/Q3	99/Q4
Organization	X			
Staff		X		
Equipment			X	
Calibration procedures				X
Test methods	X			
Reference standards			X	
Environment		X		
Handling of samples			X	
Records				X
Subcontracting				X

according to an audit schedule (Table 14.3). In the author's experience, the horizontal audit scheme is preferable.

Steps in Preparing and Conducting an Audit

Below is a step-by-step recommendation on how to prepare and conduct an audit:

Preparation

1. Establish audit team and lead auditor.
2. List areas to be evaluated (see long-term plan, if any).
3. Review results of previous audits.
4. Contact laboratory.
5. Prepare an agenda.
6. Review agenda with laboratory and reach consensus.

Conduct

1. Review selected documents, e.g., procedures.
2. Interview management.
3. Walk through the facilities, observe laboratory work and interview operators.
4. Examine test procedures and ask for some specific results. Trace the result back to methods for analysis and data

Table 14.3. Vertical audit schedule

Vertical (all items)	97/Q1	97/Q2	97/Q3	97/Q4
Lab I	X			
Lab II		X		
Lab III			X	
Lab IV				X

evaluation to equipment, operators and raw data. Review raw data. Verify if the equipment and data have been validated.

5. Give immediate advice if any noncompliance with standards has been found.

Conclusion, Report and Follow-up

1. Have a closing meeting with all auditors and laboratory management.

2. The (chief) auditor/inspector summarizes all findings, assigns level of concerns to each finding and listens to the lab's response. Any misunderstandings should be resolved at this point.

3. The (chief) auditor/inspector writes a summary report (the detailed report should not contain any surprises that were not mentioned in the summary). The chief auditor sends the report to laboratory management. This also includes a time frame when the response, e.g., 30 days, is required.

4. The laboratory resolves the problem and writes an official statement to the auditor.

5. If the statement is accepted by the audit team, the file is closed. Include the audit log that can be shown to external auditors/inspectors as proof for timely and successful audits.

Audit Report

After each audit, a report (Table 14.4) should be generated that should include

- name of the auditor(s);
- date of audit;
- areas audited;
- any noncompliance observed;
- categorization of the noncompliance (e.g., critical, serious, minor);

Table 14.4. Audit summary report template

Date	
Audit number	
Auditor(s)	
Location/Laboratory	
Details of inspections (inspected documentation, equipment, raw data, etc.) use separate sheet if necessary	
Observed nonconformities, with categories • critical • minor use separate sheet if necessary	
Recommended corrective actions	

- corrective action agreed on, responsibility for corrective action and its time frame for completion; and
- summary of audit findings, with positive statements, serious noncompliance and recommendations for corrections and improvements.

Audit Checklist

For validation activities, including calibration and testing, special attention should be paid to maintenance and change control, safety procedures, backup and recovery, error handling and recording. Table 14.5 lists possible audit items. Although there is some doubt about the usefulness of such checklists, they do, in fact, help to ensure that users have considered the most important requirements. The checklist should be used as guideline, but it is not all-inclusive.

Summary Recommendations

1. Develop a policy and procedure for audits.
2. List all audit items.
3. Develop a schedule for internal horizontal and/or vertical audits.
4. Develop checklists and templates for conducting audits.
5. Learn from audits and improve your processes.

Table 14.5. Audit questions in analytical laboratories

Audit items	Questions
Management responsibility	• Is there a documented policy and commitment to quality? • Is there a chart of the organizational structure? • Are the responsibilities of each function defined? • Are there written protocols for a QA program?
Facilities	• Is adequate space available for the type of testing performed? • Is the laboratory environment suitable for the work carried out?

Vendor assessment	• Is there a policy and procedure for purchasing equipment and chemicals?
	• Has the vendor been qualified?
	• Does the vendor have an established and maintained quality system?
	• For software vendors:
	    Does the vendor provide evidence of validation during development?
	    Are validation documents available?
	    Is the source code of software available to regulatory agencies? (This question is only important for GLP/GMP compliance.)
	• Is there an error tracking and response system for inadequate reports and enhancement requests?

Equipment	• Is there a list of all equipment used in the lab? • Are there functional and operational specifications for each piece of equipment? • Is there a protocol for IQ with test test cases, acceptance criteria and test results? • Is there a protocol for OQ with test cases, acceptance criteria and test results?

Continued on next page.

Continued from previous page

Audit items	Questions
Equipment (continued)	• Are (traceable) standards used for calibration and performance checks? • Are instruments calibrated/tested by qualified people? • Do test data sets for software qualification represent realistic data? • Have there been manual recalculations of selected critical software tasks? • Is there a preventive maintenance schedule? • Is there a schedule for ongoing calibration and PQ? • Is there a record of ongoing calibration, PQ and maintenance (logbook)? • Are instruments labeled to indicate the next OQ and/or calibration check date? • Are errors detected and recorded automatically by the system? • Are there documented procedures on error corrections? • Has defective equipment been removed from the lab or been labeled as "out of service"? • Are there documented procedures for change controls?

Test methods and procedures	• Are test methods documented?
	• Has the scope of the method been specified (criteria, performance limits)?
	• For standard (compendial) and nonstandard methods:
	    Have these methods been validated for all performance criteria, as specified by the laboratory, and are the results documented?
	    Has the suitability of such methods been verified to ensure that they are compatible with conditions that exist in the laboratory (equipment, environment, people)?

Continued on next page.

Continued from previous page

Audit items	Questions
Test methods and procedures (continued)	• For nonstandard (noncompendial) methods: Is there any documentation showing that these methods are equal to or better than standard (compendial) methods? • Does a protocol exist for changes that would require a revalidation? • Are methods periodically qualified after the initial validation? • Have alterations to the methods been authorized?
Chemical and reference material	• Are chemicals and reference materials labeled with content ID, date of acquisition/preparation and the expiration date? • Are chemicals and reference materials appropriately stored? • If refrigeration is required below a specific temperature, is the temperature monitored? • Is the reference material certified and/or traceable back to national standards? • Is the (certified) reference material obtained with a certificate? • Does the certificate state the uncertainty? • If there is no traceability, has the accuracy been otherwise verified? • Is the uncertainty of the reference material known and has this been well documented? • Has the supplier of the (certified) reference material been qualified? • Is the shelf life of the reference material known? How has it been checked and how is it documented? • Is the preparation of the working standard and reagents documented?
Samples	• Is there a procedure on sample handling? • Is there a sample tracking system? • Are samples stored appropriately?

Continued on next page.

Continued from previous page

Audit items	Questions
Documentation	• Is existing documentation (user manuals, on-line help, SOPs) adequate, complete, and up to date? • Is the documentation approved? • Does the documentation correspond to practice? • Is there an equipment logbook?
Data	• Is there an SOP for defining, collecting, entering, verifying, changing and archiving (raw) data? • Is there a procedure for checking critical data? • Where control charts are used for QC, has performance been maintained within acceptable criteria? • Is there traceability of data to equipment and people? • Is there a way to track final data back to raw data? • Do inputs or changes to data include information on who entered them, and, if they were changed, when and why?
Reporting	• Do reports provide adequate, complete and thorough information? • Have reports been dated and signed? • Do reports include information on the measurement uncertainty?
People	• Are there sufficient resources for timely response? • Are people adequately trained for their job? • Has the success of training courses been verified? • Are training records kept? • Is there an annual review of the training plan?
Internal audits	• Is there a documented procedure for inspections or audits? • Have regular internal audits been conducted?

Appendix A. Glossary

A2LA American Association for Laboratory Accreditation. A non-profit, nongovernmental, public service, membership society dedicated to the formal recognition of competent laboratories and related activities. Accredits laboratories for compliance with A2LA's accreditation standards, which includes ISO Guide 25.

AU Absorbance units.

Acceptance criteria The criteria a software product must meet to complete a test phase successfully or to achieve delivery requirements.

Accreditation The procedure by which an authoritative body gives formal recognition that a body is competent to carry out specific tasks [92].

Accuracy The degree of agreement of a measured value with the actual expected value.

AFNOR Association Francaise de Normalisation. The French Institute for Standardization.

ALADI Asociación Latinoamericana de Integración. An organization to improve commercial relations with Latin America. The members are Mexico, Argentina, Uruguay, Brazil, Bolivia, Chile, Colombia, Ecuador, Paraguay and Venezuela.

alpha (α)-test A verification test performed on a newly implemented system that mimics typical operation and that is performed by company personnel who are not system developers.

ANSI American National Standards Institute. Official standards body representing the United States with the International Organization for Standardization.

AOAC Association of Official Analytical Chemists. The primary objective of the AOAC is the development and publication of analytical methods for substances affecting public health and

	safety, economic protection of the consumer or quality of environment.
Application software	A program adapted or tailored to the specific requirements of the user for the purpose of data acquisition, data manipulation, data archiving or process control.
AQC	Analytical Quality Control.
Assay	To provide an exact result that allows an accurate statement on the content or potency of the analyte in a sample (ICH).
ASTM	American Society for Testing and Materials. A scientific and technical organization designated to develop standards on the characteristics and performance of materials, products, systems and services.
Audit	An activity to determine through investigation the adequacy of and adherence to established procedures, instructions, specifications, codes, standards or other applicable contractual and licensing requirements and the effectiveness of implementation.
Audit tracking	A procedural formality built into the operation of a system that ensures all interactions with the system are first authorized before being carried out and then recorded permanently in an operations log.
ASQ	American Society for Quality.
BCR	Bureau Communautaire de Reference. Community Bureau of Reference, Commission of the European Community, that provides certified reference material.
beta (β)-test	A verification test performed at a later stage in implementation, after debugging of the alpha-test version and at the customer's site.
Black box testing	A system/software test methodology that is derived from external specifications and requirements of the system. Methods for black box testing include random testing, testing at boundary values and a possible error list. It verifies the end results at the system level but does not check on how the system is implemented. It does not assume that all statements in the program are executed. *See also* **functional testing** [93].
British National Formulary (BNF)	Guidance on prescribing and notes on drugs and preparations, published jointly by the British Medical Association and the Royal Pharmaceutical Society of Great Britain.

British Pharmacopoeia (BP)	Official compendium of monographs providing authoritative standards for the quality of many substances, preparations and articles used in medicine and pharmacy. It incorporates monographs of the European Pharmacopoeia. It is a legally enforceable document throughout most of the Commonwealth and many other countries.
Bug	A manifestation of an error in software.
BSI	British Standards Institution.
Calibration	1. The set of operations that establish, under specified conditions, the relationship between values indicated by a measuring instrument or measuring system, or values represented by material measure and the corresponding values of the measurand. Used by regulatory agencies to refer to the process of checking or adjusting instruments (including analytical instruments). Also used in chromatography to refer to the process of using standard samples as part of method verification.
	2. An operational check that generally involves the use of standard materials or test instruments that have certification traceable to the National Institute of Standards and Technology (formerly the National Bureau of Standards).
CANDA	Computer-Assisted New Drug Application.
Change control	A procedural formality required for validation, defining how and when changes may be made and in which situations revalidation is required.
CE	Capillary electrophoresis.
CEN	Comité Européen de Normalisation. The committee on European standardization. Its members are the national standards organizations of EC and EFTA countries.
CEN/CENELEC	Comité Européen de Normalisation/Electrotechnical Standardization: The joint European Standards Institution. Develops norms such as the EN 45000 series.
Certified reference material (CRM)	Reference material, accompanied by a certificate, one or more of whose property values are certified by a procedure that establishes its traceability to an accurate realization of the unit in which the property values are expressed, and for which each certified value is accompanied by an uncertainty at a stated level of confidence (ISO Guide 30:1992).

CITAC	Co-operation on International Traceability in Analytical Chemistry. A forum for worldwide cooperation amd collaboration on the mechanisms needed to ensure the validity and comparability of analytical data on a global basis.
Certification	Procedure by which a third party gives written assurance that a product, process or service conforms to specified requirements [92].
	1. Documented review and approval of all qualification and validation documentation prior to release of the design production.
	2. Documented review and approval process performed as the final step in a validation program to permit product release.
	3. Requirement that each manufacturer of an electronic product certify that it conforms to all applicable standards.
cGMP	Current Good Manufacturing Practice.
Checksum	Programming terminology for an arithmetic operation performed on the data immediately after being generated, the product of which is stored with the data. Future access to the data is subject to the same arithmetic check. Numerical matches confirm the data have not been tampered with, while mismatches draw attention to possible corruption of the data.
Code of Federal Regulations (CFR)	Collection of all regulations issued by U.S. government agencies. The individual titles making up the regulations are numbered the same way as the federal laws on the same topic. For example, the Federal Food, Drug, and Cosmetic Act is found in Title 21 of the U.S. Code and the companion regulations implementing the law are found in 21 CFR.
COMAR	Code d'Indexation des Materiaux de Reference. International database for registering reference material. Joint enterprise between the Laboratoire National d'Essais (Paris, France), the Bundesanstalt fuer Materialforschung und Pruefung (Berlin, Germany) and the National Physical Laboratory (Teddington, United Kingdom).
Compliance	A state of laboratory operations that ensures activities follow documented protocols. GLP compliance is the responsibility of the *study director* who oversees the facility, the personnel, the materials and the equipment or subcontractors that fall under the compliance protocols. A particular instrument is

only GLP compliant when *validated* and *verified* by the operator for the specific analysis to be performed. A vendor cannot claim GLP compliance for its products.

Computer system | A system composed of computer(s); peripheral equipment, such as disks, printers and terminals; and the software necessary to make them operate together (ANSI/IEEE Standard 729-1983).

Computerized system | A system that has a computer as a major, integral part. The system is dependent on the computer software to function. [93].

Computer-related system | Computerized system plus its operating environment.

Conformity | Fulfillment by a product, process or service of specified requirements (EN 45020) [92].

Control charts | Routine charting of data obtained from the analysis of standards or (certified) reference material to check that the results lie within predetermined limits.

CSVC | PMA's Computer System Validation Committee.

DAB | Deutsches Arzneimittelbuch, German equivalent of the USP (United States Pharmacopeia).

DAD | (UV/visible) diode array detector (HPLC).

Data validation | A process used to determine if data are inaccurate, incomplete, or unreasonable. The process may include format checks, completeness, checks, check key tests, reasonableness checks and limit checks.

Debugging | The activity of first determining the exact nature and location of the suspected error within the program and second fixing or repairing the error.

Declaration of System Validation | A Hewlett-Packard publication that testifies that the HPLC ChemStation has been validated during its development and according to the Hewlett-Packard Analytical Products Group Life Cycle.

Declaration of conformity | A Hewlett-Packard publication that testifies that the equipment has been tested and found to meet shipment release specifications.

Design | The process of defining the architecture, components, interfaces, and other characteristics of an (automated) system or component.

Design Qualification (DQ)	Defines the functional and operational specifications of the instrument and details the conscious decisions in the selection of the supplier [15].
Design review	Planned, scheduled and documented audit of all pertinent aspects of the design that can affect performance, safety or effectiveness.
Design specifications	Description of the physical and functional requirements for an article. In its initial form, the design specification is a statement of functional requirements, with only general coverage of physical and test requirements. The design specification evolves through the research and development phase to reflect progressive refinements in performance, design, configuration and test requirements.
DHSS	Department of Health and Social Security. Former name of the British Health Authority, now the Department of Health.
DIN	1. Drug Information Number, an approval number given by the Canadian Health Protection Branch.

2. Deutsches Institut für Normung, the German Standards Institute. |
Disaster recovery plan	A document that lists all activities required to restore a system to the conditions that prevailed before the disaster occurred, for example, after a power failure.
EA	European Co-operation for Accreditation. Merged in November 1997 from the European Accreditation of Certification (EAC) and European Co-operation for the Accreditation of Laboratories (EAL). A major role for the EA is to develop, evaluate and ensure the maintenance of conformity assessment bodies. Membership in the EA consists of the nationally recognized accreditation bodies of the European Union and EFTA. Other non-EU/EFTA nations, with nationally recognized accreditation functions, in line with international standards, may also join as associate members.
EAC	European Accreditation of Certification (EAC). Merged in 1997 with the EAL into the EA.
EAL	European Co-operation for the Accreditation of Laboratories (EAL). Merged in 1997 with the EAC to form the EA.
EC	European Community. *See also* **EU**.

EN 45001	*General Criteria for the Operation of Testing Laboratories.* A European standard specifically intended for operating testing and calibration laboratories. EN 45001 is typically used as a guide against which a laboratory's quality system can be evaluated for accreditation. The content of EN 45001 is similar to ISO/IEC Guide 25.
EPA	Environmental Protection Agency of the U.S. government. A regulatory body that develops and enforces all aspects of environmental monitoring, including the development of analytical methods.
Equipment	Defined as the analytical measurement hardware including the firmware, for example, a gas chromatograph. In a computerized system, the equipment is controlled by the computer system. The computer system collects measurement data from the equipment.
Equipment Qualification (EQ)	The overall process of ensuring that an instrument is appropriate for its intended use.
Error	The difference between an individual result and the true value of the quantity being measured.
Escrow	Ancient legal term also applied to the deposit of source code by the software developer with an independent third party called the escrow agent. The agent holds the source code according to the terms and conditions set out in the escrow agreement, which allows the agent to release it to specified users of the software in certain circumstances. These circumstances are usually the bankruptcy or liquidation of the software developer, or the failure by the developer to carry out any maintenance obligation under the license agreement.
European Organization of Testing and Certification (EOTC)	Formed in 1990 through a Memorandum of Understanding (MOU) between the EC/EFTA members and CEN/CENELEC. Their goal is to harmonize certification and accreditation in Western Europe (EC and EFTA) in nonregulated product areas for mutual recognition of standards and accreditation throughout the EC and EFTA countries.
European Pharmacopeia	Official compendium of the member states of the Council of Europe, which includes all EC and EFTA countries.
European Union (EU)	Members in 1998 are Austria, Belgium, Denmark, Finland, France, Germany, Greece, Ireland, Italy, Luxembourg, the

Netherlands, Portugal, Spain, Sweden and the United Kingdom. Formerly called European Community (EC) and European Economic Community (EEC).

External audit Also known as a third-party audit. A periodic process carried out by an external body, e.g., the National Measurement Accreditation Service (NAMAS), to check that the laboratory's quality assurance system is effective, documented and adhered to by all staff.

External reference specifications (ERS) A Hewlett-Packard document that lists the requirements a new product under development is expected to fulfill.

FDA Food and Drug Administration, a U.S. agency part of the Department of Health and Human Services, responsible for regulating clinical research and approval of marketing permits for food, drugs, medical devices and cosmetics in the United States.

FDA compliance policy guide FDA manual for FDA field operations personnel that contains policy guidance on FDA interpretations of regulations and other compliance policies.

FDA guidelines Document published by U.S. Food and Drug Administration to provide drug sponsors with informal guidance on specific FDA requirements. Unlike regulations, guidelines are not legally binding.

FDA Inspectors Technical Guide Guide published by the Food and Drug Administration to its field inspectors. It is intended as a vehicle for making all FDA inspectors aware of selected technical information not previously available on a broad scale. Some of the topics addressed are lyophilization of parenterals, measurement of relative humidity in the ethylene oxide process, evaluation of production cleaning processes for electronic medical devices, bacterial endotoxins, new equipment static mixers, diathermy and ethylene oxide sterilization.

FDA recommendations A means the U.S. Food and Drug Administration uses to disseminate information about matters authorized by, but not involving, direct regulatory action under the laws it administers. Recommendations are not legally binding on a manufacturer.

Firmware The combination of a hardware device, e.g., an integrated circuit, and computer instructions and data that reside as

	read-only software on that device. Such software cannot be modified by the computer during processing.
Functional specification	A written definition of the function that a system or system component can perform.
Functional testing	Also known as black box testing because source code is not needed. Involves inputting normal and abnormal test cases, then evaluating outputs against those expected. Can apply to computer software or total system.
GALP	Good Automated Laboratory Practice.
GAMP	Good Automated Manufacturing Practice.
GAP	Good Analytical Practice.
GCP	Good Clinical Practice.
GMP	Good Manufacturing Practice.
Good Clinical Practices (GCPs)	Term used to describe a collection of loosely related regulations that define the responsibilities of those involved in a clinical trial. The regulations include those that govern institutional review boards, informed consent, and sponsors and monitors. Refer to 21 CFR Parts 50 and 56.
Good Laboratory Practices (GLPs)	Regulations of the U.S. Food and Drug Administration and other countries that spell out the requirements for nonclinical (animal or laboratory) studies that will be submitted to the regulatory agency to support a marketing application. The U.S. GLPs are found in 21 CFR Part 58.
Good Manufacturing Practices (GMPs)	1. Also known as current good manufacturing practices, cGMPs and GMPs. U.S. regulations in 21 CFR Part 211 contain the minimum cGMPs for methods, facilities and controls to be used for the manufacture, processing, packing or holding of a drug to assure that it meets the requirements of the Federal Food, Drug, and Cosmetic Act for safely and has the identity and strength and meets the quality and purity characteristics that it claims. Good manufacturing practices for medical devices are found in 21 CFR Part 820 and for blood and blood products in 21 CFR 606. 2. European Community Guide to Good Manufacturing Practice for Medicinal Products is the fourth volume of the *Rules Governing Medicinal Products in the European Community.* It is an EC Guide approved by representatives of the pharmaceutical inspection services of the member states of the EC and has been effective since 1 January 1992.

3. The phrase is used generally for rules, regulations or guidelines on the subject issued by any government.

ICH International Conference on Harmonization of Technical Requirements for Registration of Pharmaceuticals for Human Use. Members are from industry and regulatory agencies.

1. Provides a forum for a constructive dialog between regulatory authorities and the pharmaceutical industry on the real and perceived differences in the technical requirements for product registration in the EU, United States and Japan.

2. Identifies areas where modifications in technical requirements or greater mutual acceptance of research and development procedures could lead to a more economical use of human, animal and material resources, without compromising safety.

3. Makes recommendations on practical ways to achieve greater harmonization in the interpretation and applications of technical and requirements for registration.

ILAC International Laboratory Accreditation Cooperation. Working for international acceptance of data generated by accredited organizations. Developed the ISO Guide 25.

Interlaboratory test comparisions Organization, performance and evaluation of tests on the same or similar items or materials by two or more laboratories in accordance with predetermined conditions [92].

Internal audit A periodic process carried out by laboratory staff to check that the laboratory's quality assurance system is effective, documented and adhered to by all staff.

International standard Standard that is adopted by an international standardizing/ standards organization and made available to the public [92].

Inspection Structured peer reviews of user requirement specifications, design specifications and documentation.

Installation Qualification (IQ)
1. Installation qualification establishes that the instrument is delivered as designed and specified, that it is properly installed in the selected environment and that this environment is suitable for the operation and use of the instrument [15].

2. Documented verification that all key aspects of hardware installation adhere to appropriate codes and approved

	design intentions and that the recommendations of the manufacturer have been suitably considered [94,95].
ISO	International Organization for Standardization. Agency responsible for developing international standards; over 160 technical committees, 650 subcommittees and 1,500 working groups; more than 6,000 ISO standards published; represents more than 90 countries. Founded in 1947.
ISO 9000 series standards	The ISO 9000 series standards (9001, 9002, 9003) apply internationally. They are relevant not just for laboratories but for all types of manufacturing and service organizations. At this time, the ISO 9000 standards are primarily voluntary. However, many companies are finding that their customers are demanding ISO 9000 series compliance or registration as a condition for doing business.
ISO/IEC Guide 25	*General Requirements for the Competence of Calibration and Testing Laboratories*. Like the ISO 9000 series standards, compliance with ISO/IEC Guide 25 is voluntary. It is specifically intended only for calibration and testing laboratories. ISO/IEC Guide 25 is typically used as a guide against which a laboratory's quality system can be evaluated. Unlike the ISO 9000 series standards, it is not possible to apply for an ISO/IEC Guide 25 registration. Content of ISO/IEC Guide 25 is similar to EN 45001.
Japanese Pharmacopeia (JP)	Official pharmacopoeia of Japan.
(Laboratory) accreditation	Formal recognition that a testing laboratory is competent to carry out specific tests or types of tests (EN 45001) [36].
LGC	Laboratory of the Government Chemist, Teddington, UK.
LIMS	Laboratory Information Management System.
LOD	Limit of detection.
LOQ	Limit of quantification.
MOH	Ministry of Health, the most commonplace designation for a country's health regulation authority.
NAMAS	National Measurement and Accreditation System in the United Kingdom.
National standard	A standard that is adopted by a national standards body and made available to the public [92].

Newsgroups	Discussion groups, for example, on the Internet, that anyone can join, reading and posting articles in a worldwide forum. In 1997, there were about 10,000 discussion groups on the Internet discussing almost any imaginable topic.
NIST	National Institute for Standards and Technology in the United States. Formerly called the National Bureau of Standards (NBS). Responsible for establishing a measurement foundation to facilitate both national and international commerce.
NTIS	National Technical Information Service. Part of the U.S. Department of Commerce with responsibility for publishing and sales of technical documents. It has similar publishing functions as the U.S. Government Printing Office.
NVLAP	National Voluntary Laboratory Accreditation Program. A federal program under which NVLAP operates as an unbiased third party to accredit both calibration and testing laboratories (http://ts.nist.gov/nvlap).
Obsolescence	The final phase in a system's life cycle when the system is retired from use and taken off the market. At Hewlett-Packard, an obsolescence plan documents the support activities guaranteed for up to 10 years following obsolescence.
OECD	Organization for Economic Co-operation and Development.
Operational Qualification (OQ)	1. Process of demonstrating that an instrument will function according to its operational specifications in the selected environment [15].
	2. Documented verification that the equipment-related system or subsystem performs as intended throughout representative or anticipated operating ranges [94, 95].
Out-of-control	Reference to a situation in which compliance with GMPs is not evident. The facility or operation is considered to be out of control.
PASG	(UK) Pharmaceutical Analytical Sciences Group.
Password	In computers and computer software, a security identification text only known to authorized operators, often with capability levels.
Performance Qualification (PQ)	1. Process of demonstrating that an instrument consistently performs according to a specification appropriate for its routine use [15].

2. Documented verification that the process and/or the total process related system performs as intended throughout all anticipated operating ranges.

Performance Verification (PV) A service offered by Hewlett-Packard's Analytical Products Group support organization. It verifies that the system at the user's site performs according to the specifications as agreed between the vendor and the purchaser. Chromatographic performance specifications are published in the vendor's specification sheets. Chromatographic instrument hardware specifications include baseline noise and precision of retention times and peak areas.

Pharmacopeia Official compilation of medicinal substances and/or articles with descriptions, tests and formulas for preparing them, selected by a recognized authority. The pharmacopoeia issued for a country is the legal standard of that nation. Also spelled **pharmacopoeia.**

PIC Pharmaceutical Inspection Convention, a multinational organization (primarily of European countries) whose members have agreed to mutual recognition of facility inspections for good manufacturing practice.

PICSVF UK Pharmaceutical Industry Computer Systems Validation Forum.

PMA Pharmaceutical Manufacturers Association in the United States. A trade association that represents more than 100 firms, collectively producing more than 90 percent of American prescription drugs. Now known as the Pharmaceutical Research and Manufacturers of America (PhRMA).

Precision The degree of agreement of a measured value with other values recorded at the same time, in the same place or on similar instruments. Also referred to as repeatability.

Proficiency testing Determination of laboratory testing performance by means of interlaboratory test comparisons [92].

A systematic testing program in which samples are analyzed by a number of laboratories to measure the competence to undertake certain analyses.

Prospective validation Establishing documented evidence that a system does what it purports to do based on a validation plan [33].

Prototyping	An approach to accelerate the software development process by facilitating the identification of required functionality during analysis and design phases. A limitation of this technique is the identification of system and software problems and hazards.
Qualification	Action of proving that any equipment works correctly and actually leads to the expected results. The word *validation* is sometimes widened to incorporate the concept of qualification.
Quality assurance	A set of activities, often performed by employees in a similarly named department, that checks that the characteristics or qualities of a product actually exist at the time the product is sold. Oversight function that audits operations to determine that procedures and systems are suitable and recommends required changes to provide evidence that the quality function is functioning correctly. Quality assurance is involved from product concept through design, manufacture and distribution until the ultimate use of the product by the patient.
Quality control	Day-to-day control of quality within a company, responsible for the acceptance or rejection of incoming raw materials and packaging components, in-process tests, labeling and inspection; assurance that systems are being controlled and monitored and for the approval or rejection of finished dosage forms. A laboratory-based function.
Raw data	Any laboratory worksheets, records, memoranda, notes or exact copies thereof that are the result of original observations and activities of a nonclinical laboratory study and are necessary for the reconstruction and evaluation of the report of that study. It may include photographs, microfilm or microfiche copies, computer printouts, magnetic media, including dictated observations, and recorded data from automated instruments.
Reference material (RM)	A material or substance, one or more properties of which are sufficiently well established to be used for calibrating an apparatus, assessing a measurement method or for assigning values to materials [37].
Reference standard	A standard, generally of the highest metrological quality available at a given location from which measurements made at that location are derived [59].

Regulatory methods validation	Process whereby submitted analytical procedures are first reviewed for adequacy and completeness and then are tested as deemed necessary in U.S. Food and Drug Administration laboratories. Depending in part on the quality of submitted data, validation may range from step-by-step repetition of an assay procedure to more elaborate studies that include assessment of accuracy, precision, sensitivity and ruggedness of the method.
REMCO	Council Committee of Reference Materials of the International Organization for Standardization, established in 1976. The committee has since published several guides on the nomenclature, certification and uses of reference materials.
Retrospective validation	Establishing documented evidence that a system does what it purports to do based on review and analysis of historic information.
Revalidation	A repetition of validation necessary after the process has been changed, for example, when a manual system is upgraded to an automated system [33].
Ruggedness	An indication of how resistant the process is to typical variations in operation, such as those to be expected when using different analysts, different instruments and different reagent lots. Required under GLP guidelines.
SI	System International.
Source code	An original computer program in a legible form (programming language), translated into machine-readable form for execution by the computer.
Standard Operating Procedure (SOP)	Documented instructions that should be followed when operating a process for the process to be considered valid. Required under GLP regulations. Written documents that prescribe the detailed methods and action steps to be followed in order to accomplish a particular task. The U.S. Food and Drug Administration requires SOPs for virtually every aspect of production, control and testing of pharmaceutical products. One of the SOPs should describe the issuance and control of SOPs. The UK Guide to Good Pharmaceutical Manufacturing Practices defines SOPs as written, authorized procedures that give instructions for performing operations not necessarily

specific to a given product or material, but of a more general nature (equipment operation, maintenance and cleaning, cleaning of premises and environmental control, sampling and inspection, etc.).

System suitability testing A process of checking out the performance specifications of a system, often called method validation when applied to a particular separation and called system validation when applied to a separation system used routinely.

TGA (Australian) Therapeutic Goods Administration.

Test A technical operation that consists of the determination of one or more characteristics or performance of a given product, material, equipment, organism, physical phenomenon, process or service according to a specified procedure [37].

Test plan A document prescribing the approach to be taken for intended testing activities. The plan typically identifies the items to be tested, the testing to be performed, test schedules, personnel requirements, reporting requirements, evaluation criteria and any risks requiring contingency planning.

Traceability The property of a result of a measurement whereby it can be related to appropriate standards, generally international or national standards, through an unbroken chain of comparisons all having stated uncertainties [37].

UKAS United Kingdom Accreditation Service. The national accreditation body for the United Kingdom formed in 1995 by the amalgamation of the National Measurement and Accreditation System (NAMAS) and the National Accreditation Council for Certification Bodies (NACCB).

Uncertainty Measurement uncertainty is an estimate of a measurement that characterizes the range of values within which the true value is asserted to lie (ISO/DIS 254-1).

URL Uniform resource locator. Tool used to identify sites on the Internet.

USP U.S. Pharmacopeia (USP). Official compendium recognized by the Federal Food, Drug, and Cosmetic Act. Serves as the basis for enforcement actions by the U.S. Food and Drug Administration involving official (USP) drugs. Published every five years by the United States Pharmacopeial Convention, a nonprofit organization. It is combined with the National

	Formulary. The USP is the official pharmacopeia of the United States and several other countries.
Validation	Establishing documented evidence that provides a high degree of assurance that a specific process will consistently produce a product meeting its predetermined specifications and quality attributes [43].
Validation protocol	Written plan stating how validation will be conducted, including test parameters, product characteristics, production equipment and decision points on what constitutes acceptable test results.
Verification	Confirmation by examination and provision of evidence that specified requirements have been met [37].
Warning Letter	Letter issued by U.S. Food and Drug Administration to a manufacturer containing adverse findings and giving the manufacturer 15 days in which to reply. It replaced the Regulatory Letter and the Notice of Adverse Findings.
Western European Laboratory Accreditation Conference (WELAC)	Formed in 1989 to represent the interests of laboratory accreditation bodies in Western Europe. Deals with accreditation of test laboratories for European recognition by their clients. Established through a Memorandum of Understanding (MOU) for multilateral recognition of accredited laboratories. Evaluates accreditation and certification bodies for European recognition of test laboratories accredited by the accreditation body. Also has contacts with the OECD on the relationship between GLP and EN 45001.
	Audits accreditation bodies in Europe, for example, the EAM in Switzerland.
	Combined with the WECC in 1994. Developed and signed a contract with the EOTC for EC–wide recognition of accreditation systems (May 13, 1992).
WWW	World Wide Web. Collection of pages that can be published by anyone and can be viewed by millions of Internet users. The Web is the most popular method of distributing information through the Internet.

Appendix B. OQ Tests for Selected Equipment

This appendix summarizes test procedures and acceptance limits for the OQ of selected equipment in a table format. A few general recommendations that apply to all procedures are given first.

Traceability of Standards

Typically, standards used for OQ performance tests should be certified or traceable to national standards. Test samples not used for quantitative analysis do not need to be certified or traceable to national standards. An example is a sample that is used to determine the precision of peak area or peak retention time. Samples that are used for quantitative calibration of the system should be certified and/or traceable to a national standard. An example is a standard that is used to determine the linearity of a detector. In those cases where external certification or traceability for such samples is not available, the laboratory should do the utmost to ensure accuracy of the standard. The procedure to ensure accuracy should be documented.

Acceptance Limits

Acceptance limits are determined by the intended use and may be more stringent than the ones that are recommended in the tables.

Documentation and Archiving

In a summary report, the acceptance limits and the actual results should be documented. The report should also include the date of measurement and the name of the operator. The report should also have a reference to the instrument identification and measurement method.

If the actual results have been calculated from a series of measurements, the results of the individual measurements should be archived for later recalculation, if necessary. For example, when the standard deviation of an injection precision of a gas chromatography autosampler is calculated from six replicate injections, the peak area of individual measurements should be recorded and archived.

Gas Chromatography

Precision of peak retention times	
Test procedure	1. Five injections of a standard
	2. Calculation of RSD
Acceptance limits	< 1% RSD
Test frequency	Yearly
Remarks	

Precision of peak areas	
Test procedure	1. Five injections of a standard
	2. Calculation of RSD
Acceptance limits	< 2% RSD
Test frequency	Yearly
Remarks	

Accuracy of the temperature of the column oven

Test procedure	Measure temperature in column oven and compare with setpoint.
Acceptance limits	±1°C
Test frequency	Yearly
Remarks	The temperature measurement device should be calibrated and traceable to a national standard.

Capillary Electrophoresis

Stability of voltage

Test procedure	Plot of voltage
Acceptance limits	< 0.25 kV
Test frequency	Every 6 months
Remarks	

Precision of peak areas

Test procedure	1. Five injections of a standard 2. Calculation of RSD
Acceptance limits	< 5% RSD
Test frequency	Yearly
Remarks	

UV/Visible Spectrophotometer

Wavelength accuracy—holmium oxide solution

Test procedure	Measure absorption maxima of holmium oxide solution at 241.15, 287.15, 361.5 and 536.3 nm.
Acceptance limits	±1 nm in UV range ±3 nm in visible range
Test frequency	Every 6 months
Remarks	Standard should be traceable to national standard.

Wavelength accuracy—holmium oxide filter

Test procedure	Measure absorption maxima of holmium oxide filter at 361, 418 and 536 nm.
Acceptance limits	±1 nm in UV range ±3 nm in visible range
Test frequency	Every 6 months
Remarks	Filter should be traceable to national standard.

Photometric accuracy—absorbance

Test procedure	Measure absorption of potassium chromate solution.		
Acceptance limits	**wavelength**	**A 1%/1 cm**	**limits**
	235 (min)	124.5	122.9–126.2
	257 (max)	144.0	142.4–145.7
	313 (min)	48.6	47.0–50.3
	350 (max)	106.6	104.9–108.2
Test frequency	Every 6 months		
Remarks	Standard should be traceable to national standard.		

Stray light—potassium chloride

Test procedure	Measure absorption of 1.2% potassium chloride solution at 200 nm against water.
Acceptance limits	Absorption at 200 nm > 2.0 AU
Test frequency	Every 6 months
Remarks	Potassium chloride should be traceable to national standard.

Control of cuvettes

Test procedure	Transmission of cell (water) against air

Acceptance limits	a) Quartz cells
	   85% at 220 nm
	   88% at 240 nm
	
	b) Glass cells
	   85% at 356 nm
	   88% at 650 nm

Test frequency	Every 6 months
Remarks	

Wavelength resolution—0.02% toluene in hexane

Test procedure	Measure toluene spectrum from 260 to 275 nm. Calculate ratio of peak height to valley.
Acceptance limits	Absorbance ratio peak to valley at 266/269 nm > 1.5
Test frequency	Every 6 months
Remarks	DAB and EP (European Pharmacopoeia)

High Performance Liquid Chromatography

Accuracy of the flow rate

Test procedure	Measure flow rate with a volumetric flask and stopwatch or with calibrated digital flowmeter.
Acceptance limits	±5°C
Test frequency	Yearly
Remarks	This test typically is performed at the beginning of the system test procedures. Successful completion proves that there is no major leak in the system.

Precision of flow rate

Test procedure	1. Five injections of a standard 2. Calculation of RSD
Acceptance limits	< 2% RSD
Test frequency	Yearly
Remarks	Recommended standard: caffeine

Precision of peak areas

Test procedure	1. Five injections of a standard 2. Calculation of RSD
Acceptance limits	< 2% RSD
Test frequency	Yearly
Remarks	Recommended standard: caffeine

Accuracy of the temperature of the column oven

Test procedure	Measure temperature in column oven and compare with setpoint.
Acceptance limits	±1°C
Test frequency	Yearly
Remarks	1. Column temperature accuracy is important when methods are transferred between instruments. 2. The temperature measurement device should be calibrated and traceable to a national standard.

Precision of the temperature of the column oven

Test procedure	Measure temperature in column over 20 min at 40°C.
Acceptance limits	±0.5°C
Test frequency	Yearly
Remarks	

Baseline noise

Test procedure	Plot baseline for 20 min. Measure peak-to-peak noise in sections of 1 min. Average results.
Acceptance limits	$\pm 1 \times 10^{-4}$°C
Test frequency	Yearly
Remarks	ASTM E19.09

Detector linearity

Test procedure	Inject 5 standards of caffeine. Plot response factor versus amount.
Acceptance limits	≤1.5 AU with 5% level.
Test frequency	Yearly
Remarks	Standards should be certified (available from Hewlett-Packard).

UV/visible detector wavelength accuracy

Test procedure	Scan compound with known spectrum. Measure wavelength at absorption maximum. Compare actual value with reference value.
Acceptance limits	±2 nm
Test frequency	Every 3 months and whenever the detector is moved.
Remarks	1. Known standard should be certified.
	2. Caffeine is a good standard for 250 to 300 nm.
	3. Built-in holmium oxide filters can be used for automated checks.

Autosampler carryover

Test procedure	Inject blank solvent after standard. Measure peak ratio between blank and standard injection.
Acceptance limits	< 0.3%
Test frequency	Yearly
Remarks	Caffeine has been proven to be a good standard.

Mobile phase composition accuracy

Test procedure	Run step gradients at 10, 11, 50 and at 90% with acetone tracer; step heights relative to 100%.
Acceptance limits	±2%
Test frequency	Yearly
Remarks	The analytical column should be replaced by empty tubing. Back pressure should be > 20 bar.

Infrared/Near Infrared

Wavelength accuracy

Test procedure	Measure polystyrene spectrum at 1144, 1680, 2167 and 2307 nm. Compare results with reference values.
Acceptance limits	±2 nm
Test frequency	Daily
Remarks	Standard should be traceable to national standard.

Control at 0 and 100%

Test procedure	Control at 0 and 100%
Acceptance limits	1% transmission (measured at 4,000 cm^{-1})
Test frequency	Daily and after changing measurement parameters
Remarks	

Wavelength resolution

Test procedure	Resolution of polystyrene at 2870/2851 and at 1589/1883 nm
Acceptance limits	ΔT (band 2870 cm^{-1}–band 2851 cm^{-1}) ≥ 18 ΔT (band 1589 cm^{-1}–band 1583 cm^{-1}) ≥ 12
Test frequency	Daily
Remarks	

Analytical Balance

Accuracy	
Test procedure	Measurement of reference weight, use 10 mg, 50 mg, 100 mg, 500 mg, 1 g, 5 g, 10 g and 20 g. Compare the actual results with reference weights.
Acceptance limits	0.1%
Test frequency	Daily or when used, whichever is longer, with internal reference weights
	Yearly with traceable external weights through instrument vendor
Remarks	External standard should be traceable to national standard.

Flame Atomic Absorption Spectrophotometer

Calibration	
Test procedure	As stated in analytical method
Acceptance limits	See vendor specifications
Test frequency	Every 6 months
Remarks	Standard as stated in analytical method

Linearity	
Test procedure	As stated in analytical method
Acceptance limits	See vendor specifications
Test frequency	Daily
Remarks	Standard as stated in analytical method as defined in method validation

Laboratory Ovens

Temperature accuracy

Test procedure	Measure the temperature inside the oven over the full temperature working range using at least 6 different temperatures and compare with setpoints. Plot a curve of actual oven temperature versus setpoints.
Acceptance limits	±2°C. If deviations are higher, the measured data points should be plotted as a calibration curve. The curve should be used for actual temperature adjustments.
Test frequency	Every 6 months
Remarks	1. Thermometer should be calibrated and traceable to national standard.
	2. For critical applications, a calibrated thermometer should be mounted inside the oven for continuous monitoring.

Laboratory Furnaces

Temperature accuracy

Test procedure	Measure the temperature inside the furnace over the full temperature working range using at least 6 different temperatures and compare with setpoints. Plot a curve of actual furnace temperature versus setpoints.
Acceptance limits	±5°C. If deviations are higher, the measured data points should be plotted as a calibration curve. The curve should be used for actual temperature adjustments.
Test frequency	Yearly
Remarks	1. Thermometer should be calibrated and traceable to national standard.
	2. For critical applications, a calibrated thermometer should be mounted inside the oven for continuous monitoring.

Sterilizers (Hot Air)

Temperature accuracy

Test procedure	Measure the temperature inside the sterilizer using a thermometer.
	Measure the temperature inside the sterilizer at various locations using a thermocouple.
Acceptance limits	±2°C
Test frequency	Daily with thermometer
	Bimonthly with thermocouple
Remarks	Thermometer should be calibrated and traceable to national standard.

Refrigerators and Freezers

Temperature accuracy

Test procedure	Measure the temperature inside the refrigerator over the full temperature working range using at least 6 different temperatures and compare with setpoints. Selected temperature range may be 4 to 20°C. Plot a curve of actual refrigerator temperature versus setpoints.
Acceptance limits	±2°C
Test frequency	Yearly
Remarks	1. Thermometer should be calibrated and traceable to national standard.
	2. For critical applications, a calibrated thermometer should be mounted inside the refrigerator or freezer for continuous monitoring.
	3. Some accreditation schemes require daily or weekly monitoring of the temperature.

Thermometers and Thermocouples

Temperature accuracy	
Test procedure	Measure the temperature of the thermometer over the full working range at characteristic reference points, e.g., at the ice point (0°C).
Acceptance limits	±1°C at 30 to 40°C; ±2°C at 100°C
Test frequency	Yearly

Remarks	1. In addition to the specified in-house procedure the thermometer should be calibrated by external organizations (e.g., by an accredited laboratory) at least every year.
	
	2. Some applications require better accuracy than those specified under acceptance limits, e.g., the temperature of incubation for some microbiological tests could be as tight as ±0.25°C. In this case, the thermometers should be calibrated by an external service.
	
	3. Temperature reference points:
	
	    triple point of equil. hydrogen −259.34°C
	    boiling point of oxygen −182.96°C
	    melting point of water 0°C
	    boiling point of water 100.0°C
	    freezing point of zinc 419.58°C
	    freezing point of silver 961.93°C
	    freezing point of gold 1,064.43°C

Karl Fisher Apparatus

Calibration and precision

Test procedure	Add known amount of water (~ 50 mg, 1 drop) to 100 mL anhydrous methanol. Titrate the water with Karl Fisher reagent (e.g., pyridine based). Calculate the water equivalence factor f using $$\frac{\text{water (mg)}}{\text{mL KF reagent}}$$ The measurement should be performed 3 times, and the results should be averaged.
Acceptance limits	Precision < 1% RSD
Test frequency	Before each use
Remarks	

Analog/Digital Converter (from Ref. 97)

Analog input—A/D converter	
Test procedure	1. Generate a voltage of about 1 V with commercial battery and voltage separator. 2. Apply the voltage to the analog input and read the value on the data system. 3. Compare the value on the data system with the voltmeter. 4. 0 V measurement at the analog input.
Electrical diagram	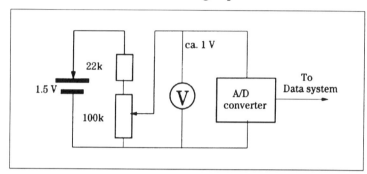
Acceptance limits	±3%
Test frequency	Yearly
Remarks	1. Use calibrated voltmeter. 2. The environmental temperature should be stable (±1°C over the time of measurement cycle). 3. There should not be any electromagnetic interference. 4. Don't touch electrical cables and voltage separator during the measurement cycle. 5. Use new battery for each measurement. 6. Remove the battery after the measurment.

Dissolution Testing

Temperature accuracy—dissolution bath

Test procedure	Measurement of temperature in test sample and water bath with a calibrated thermometer at 37°C
Acceptance limits	37°C ± 0.5°C
Test frequency	Every 6 months
Remarks	According to USP, EP and German DAB
	Thermometer should be traceable to national standard.

Accuracy of shaft rotation

Test procedure	Measure shaft rotation at 50 and 100 turns/min with rotation measurement device and stopwatch. Compare results with setpoint.
Acceptance limits	±4%
Test frequency	Every 6 months
Remarks	DAB and EP

Control of distance of shaft from side of vessels

Test procedure	Measurement of the distance of the stirrer
Acceptance limits	25 mm ± 2 mm
Test frequency	Before each measurement (according to German DAB)
Remarks	German DAB requirement is very time-consuming.

Control of paddle centering

Test procedure	Difference between the axis of the rod and the axis of the vessel
Acceptance limits	< 2 mm
Test frequency	Before each measurement
Remarks	

Control of instrument suitability

Test procedure	Measurement of the dissolution rate with salicylic acid tablets and Prednisone tablets
	Measure 1 tablet of disintegrating type and 1 tablet of nondisintegrating type.
Acceptance limits	Compound specific
Test frequency	Every month

Remarks	USP Dissolution Calibrator
	    disintegrating type, e.g., Salicylic acid tablet
	    nondisintegrating type, e.g., Prednisone tablet

Viscosimeter

Accuracy

Test procedure	Measurement with calibrated oils
Acceptance limits	Very much dependent on instrument type (see manufacturer's specifications)
Test frequency	Every 6 months
Remarks	1. According to USP and German DAB
	2. Oils should be traceable to national standard (available from vendors).

Melting Point

Accuracy

Test procedure	Measurement of melting point of reference compounds of the World Health Organization
	• Heating rate: 1°C/min • Temperature: 5°C below the melting point • Vanilline: 1.5° range from 81.0 to 83.0°C • Acetanilide: 1.0° from 114.0 to 115.5°C • Phenacetine: 1.5° from 134.0 to 136.0°C • Sulfanilamide: 1.5° from 164.6 to 165.5°C • Sulfapyridine: 1.5° from 190.2 to 192.5°C • Caffeine: 1.0° from 123.5 to 237.0°C (Ref.: Mettler FP81)
Acceptance limits	0.5°C
Test frequency	Every 3 months
Remarks	Benzoic acid is also suitable.

pH Meter

Accuracy

Test procedure	Calibrate with buffer solution at pH 7 and pH 4. Verify with buffer solution at pH 5 or pH 6.
Acceptance limits	< 0.05 pH units
Test frequency	Before each use
Remarks	1. Buffer solutions are commercially available in either tablet or solution form. 2. Some pH meters also have a temperature sensor. This should be calibrated every 6 months. 3. The electrodes should be washed between each measurement, and the buffers should be visually checked for cleanness and absence of microbial growth before each use.

Refractometer

Accuracy according to USP

Test procedure	Measurement of deionized water at 20°C ± 0.1° at λ = 589.3 nm (Frauenhofer line of sodium light). Expected refractive index: $n_D(20) = 1.3330$.
Acceptance limits	$n_D(20) = 1.3325$ to 1.3335 ±5 units of the fourth digit
Test frequency	Monthly
Remarks	USP XXXIII

Accuracy according to DAP

Test frequency	Monthly
Remarks	German DAP 1966 V 6.5

Polarimeter

Accuracy

Test procedure	Measurement of calibrated control quartz Reference value $[\alpha]_D(20) = 14.986°$
Acceptance limits	$[\alpha]_D(20) = 14.97$ to $15.00°$
Test frequency	Every 3 months
Remarks	Quartz should be traceable to national standard.

Wavelength accuracy	
Test procedure	Measurement of sacharose solution according to *Ph. Eur.* 1, Vol. 1 (1974), layer thickness = 1.00 dm
Acceptance limits	C g/100 mL $[\alpha]_D(20)$ Tolerance 20.0 13.32° ±0.027° 30.0 19.95° ±0.040° 40.0 26.56° ±0.050°
Test frequency	Every 3 months
Remarks	

Declaration of Operational Qualification

At the end of the OQ tests, a Certificate or Declaration of Operational Qualification should be developed to document the tests. An example is shown below.

Declaration of Operational Qualification

Equipment	Balance
Serial number	XZ ddd
Asset number	5634-44
Date	6.6.97
Method/SOP Number	LAB567
Results and evaluation	The result meets the requirements.
Attachments	Protocol and table with reference and actual value
	Printouts
Date of next qualification	9.6.97
Signatures	
Technician Name: L. Jones	Signature: *Lou Jones*
Manager Name: B. Miller	Signature: *Brenda Miller*

Appendix C. Selected (Standard) Operating Procedures

Operating procedures or SOPs play a major role in an analytical laboratory's quality system. They are required by all good practice regulations and quality and accreditation standards. They are developed to ensure that laboratory operations are conducted with consistently good planning, appropriate distribution and execution and complete documentation. Most laboratory audits and inspections will check if appropriate SOPs are existing and followed.

This appendix gives some recommendations on SOPs, including a list of operating procedures for analytical equipment. Examples of SOPs are provided for the validation of methods and simple and complex application software, for retrospective evaluation and validation of existing computer systems and for testing of hardware. (SOPs are available in electronic format from the author.)

General Recommendations

Organization/Format

- At first, an SOP should be developed on how to develop, test, write, publish, distribute, maintain and archive SOPs. This ensures that all SOPs are handled in the same way.

- Develop and communicate a numbering and naming system for SOPs.

- Do not prepare too many SOPs. Think twice before you prepare an SOP for a special job (too much paperwork does not improve the efficiency of a laboratory or the quality of analytical data).

- Where possible, combine individual procedures into a single, larger SOP.

- The format should be consistent for all SOPs. All SOPs should include some general elements such as a meaningful title, scope, purpose (objective), procedural text and references (if any).

- The text should describe the sequence of tasks in a step-by-step format.

- All pages should numbered, by page and the total number of pages (e.g., page 1 of 3).

Content/Level of Detail

- For equipment calibration, testing and maintenance, develop one and only one SOP for each type of equipment. It should be independent from the instrument's use and manufacturer.

- Do not use the vendor's specifications as acceptance limits for equipment testing. This may require frequent preventive maintenance to meet the stringent specifications. Acceptance limits should be based on the intended use of the equipment.

- Do not be too detailed; this avoids the need for frequent updates. For example, a SOP for purchasing chemicals should not include any vendor's name, thus avoiding an update if the vendor changes.

- Do not be too restrictive. This requires users to spend too much time writing and authoring deviations. This is especially important for acceptance limits of equipment.

- SOPs not only describe what to do but how to do it.

- Instrument SOPs should be written close to the instrument in the laboratory, not in the office. They should be either written or thoroughly reviewed by somebody who has a good understanding of the technical work. SOPs should not explain how procedures are supposed to work, but how they work in reality.

- SOPs should be written so they are understood by typical users.

- SOPs should be written in a language that is understandable by the user.
- SOPs for equipment testing should include forms and templates for entries of dates, results, comments, further actions (in case the specifications are not met) and signatures.
- Books, instrument operating manuals and other literature can be used as references. However, it should be ensured that the references will be accessible during the entire archiving period.

Development and Testing

- Draft SOPs should be circulated to the target audience prior to the test release to collect inputs.
- SOPs should be tested by typical users prior to their final release.

Approval

- SOPs should be signed by the author and approved and signed by the management. Some companies also require the QA department to review and sign SOPs before they are submitted to the management.
- Copies of equipment SOPs should be located close to the instruments for easy access by operators.
- Deviations from SOPs should be explained and authorized.

Distribution and Archiving

- The distribution of SOPs should be handled by the QA department. This ensures that the distribution list is always the same and that all users always have the same version.
- Old SOPs should be returned to the QA department or scratched.
- If SOPs are distributed electronically, the users should be notified by electronic mail that a new or revised SOP is available on a specified server.
- It is a good practice to have one copy of each actual SOP stored in paper format at a central place that is easily accessible by all users.

- There should be a list of all SOPs with titles and numbers easily accessible by users.
- SOPs must be archived.

Maintenance and Periodic Review

- SOPs should be reviewed periodically to determine if the written procedure still reflects laboratory practices.
- The person responsible for the SOP should keep a record of all reviews and changes.

Types and/or Content of SOPs

SOPs can be developed for the following:

Administration

- ☑ Responsibilities
- ☑ Development and handling (distribution, archiving, etc.) of SOPs
- ☑ Naming and numbering system for SOPs

Equipment

- ☑ Purchasing of equipment and chemicals
- ☑ Qualification of a vendor
- ☑ Software development and validation (life cycle)
- ☑ Installation and operational qualification of equipment
- ☑ Retrospective evaluation and validation of existing systems
- ☑ Routine inspection, testing, maintenance and calibration
- ☑ Actions to be taken in response to equipment failure
- ☑ Logbooks and instrument repair

Data

- ☑ Definition of raw data
- ☑ Entry of data and proper identification of individuals entering the data
- ☑ Validation and changes to data

Appendix C. Selected (Standard) Operating Procedures

Analytical Methods
- ☑ Development and validation of analytical methods
- ☑ Validation of standard methods
- ☑ Validation of ad hoc methods

Handling of Samples and Standards
- ☑ Receipt and distribution of test and control samples
- ☑ Labeling of reagents and samples
- ☑ Sample collection and tracking
- ☑ Preparing standard solutions
- ☑ Processing and analyzing specific matrices and samples

Safety
- ☑ General laboratory safety issues
- ☑ Chemical hazard handling (e.g., purchasing, classification, inventory, disposal)
- ☑ Safety of visitors (safety sheets, clothing, glasses)

Security
- ☑ Limited access to buildings, equipment and data
- ☑ Generation and distribution of passwords
- ☑ Check of computer systems for viruses
- ☑ Program, method and data backup
- ☑ Disaster recovery

Personnel
- ☑ Development and communication of job descriptions
- ☑ Training of personnel

QA Audits and reviews
- ☑ Audit master schedule
- ☑ Data review and reports
- ☑ Archiving audit reports

Archiving

- ☑ Archiving system
- ☑ Submission and retrieval
- ☑ Limited access
- ☑ Storage conditions and verification

Proposal for a Title Page

Title			
Company	Company Stamp (colored, to identify black copies)	Code (ID)	Valid from
Labor	Version	Version Replaced by this SOP	
Developed By Signature: Signature: Date:	Authorized By Signature: Signature: Date:	Distribution List — — — —	
1. Scope 2. Purpose Procedural text 			
Page 1 of X			

General Workflow of SOPs for Equipment Testing

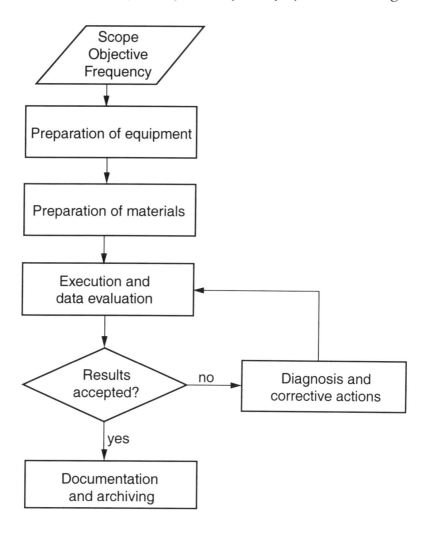

Example #1: SOP for the Preparation of Standard Operating Procedures

The development, testing, publishing and distribution of SOPs should follow documented procedures. The format, structure and some content elements should be consistent. An example is shown on the following pages. This is a proposal and starting point only and may need adaptation to different SOPs. There is no assurance expressed that the operating procedure will pass a regulatory inspection.

Company	Title: Preparation of Standard Operating Procedures	Code LHEQ11	Revision Number A.01
Division	Laboratory	Manager	Page 1 of 4
Effective Date	Prepared by Name: Sig: Date:	Approved by Name: Sig: Date:	Distribution — — — —

1. Scope

Analytical laboratories: all departments, routine procedures

2. Purpose

Regulations and quality and accreditation standards applicable to analytical laboratories require various routine laboratory activities to follow SOPs. The development, publishing, distribution, maintenance and archiving of such SOPs should follow documented procedures. The format, structure and some content elements should be consistent within a laboratory. This SOP addresses the process for the generation and maintenance of SOPs.

3. Frequency

a) When generating a new SOP
b) When revising an existing SOP

Company	Title: Preparation of Standard Operating Procedures	Code LHEQ11	Revision Number A.01
Division	Laboratory	Author	Page 2 of 4

4. Format

The SOP consists of a title page and procedural text.

a) The title page includes:
   1. The company name
   2. The title of the SOP
   3. Unique number and revision number
   4. Effective date
   5. The division and laboratory name
   6. The printed names and signatures of the author and a management representative
   7. The distribution list (by department)
   8. Page number and total number of pages
b) Each page includes:
   1. The company name
   2. The division name
   3. The laboratory name
   4. The title and ID code of the SOP
   5. The revision number
   6. The author
   7. The page number and total number of pages
c) The text part includes
   1. The scope
   2. The purpose
   3. Frequency of use
   4. Detailed procedure
   5. History of the SOP

5. Text of the SOP

a) Scope
   Defines applicability, for example, to specific departments or test procedures.

Company	Title: Preparation of Standard Operating Procedures	Code LHEQ11	Revision Number A.01
Division	Laboratory	Author	Page 3 of 4

b) Purpose
   Defines the objective, for example, to test HPLC equipment.
c) Frequency
   Defines the interval at which the SOP will be applied. Examples are daily, monthly, yearly or when used.
d) Detailed Procedure
   1. Material used; for example, chemicals
   2. Equipment used, for example, traceable test equipment
   3. Step-by-step instructions, for example, to calibrate a balance
   4. Evaluation procedure, if needed
   5. Acceptance procedure
   6. Documentation
e) History
   Defines dates of first release and dates of all consecutive revisions.

6. Preparation

a) Confirm the need for the new or revised SOP.
b) Create a draft.
c) Submit the draft to the management representative.
d) The management distributes the draft to the staff for review and comments.
e) Comments are sent to the author.
f) The edited SOP is submitted to the management for review.
g) The edited SOP is submitted by management to staff for review and testing, if appropriate.
h) The final version is approved and signed by management.

Company	Title: Preparation of Standard Operating Procedures	Code LHEQ11	Revision Number A.01
Division	Laboratory	Author	Page 4 of 4

7. Distribution

a) The management representative assigns a SOP responsible person.
b) The responsible person distributes paper copies to the QA department and to the laboratory manager.
c) The laboratory manager informs the responsible person on how many copies are required for the lab.
d) The responsible person creates the specified number of SOPs and distributes them.
e) If the newly distributed version is a revision of an older one, the older one should be sent back to the responsible person and discharged.
f) One copy is sent to the archive. The SOP archive retains all original SOPs, including revisions, in a historical file.
g) The SOP is entered into the company's electronic SOP database.

8. Maintenance, Review and Update (Change Control)

a) Every 12 months, the author or a person designated by management reviews the SOP and revises it, if required.
b) SOPs may also be revised more frequently, if necessary. (Users of an SOP are encouraged to give feedback and to make enhancement requests as part of the laboratory's quality policy).
c) Revisions follow Sections 5, 6 and 7 of this SOP.

Example #2: Validation of Analytical Methods

Analytical methods should be validated prior to routine use. An example is shown on the following pages. This is a proposal and starting point only and may need adaptation to different SOPs. There is no assurance expressed that the operating procedure will pass a regulatory inspection.

Company	Title: Validation of Analytical Methods	Code	Revision Number
		LHEQ12	A.01
Division	Laboratory	Manager	Page 1 of 9
Effective Date	Prepared By Name: Sig: Date:	Approved By Name: Sig: Date:	Distribution — — — — —

1. Scope

a) Analytical chromatographic routine methods that are developed in-house
b) Standard chromatographic methods

2. Purpose

Accurate and consistent analytical data can only be obtained with validated methods. Regulations and quality and accreditation standards applicable to analytical laboratories also require analytical methods to be validated prior to routine use and revalidated after a change. This SOP addresses the process for the validation of analytical methods.

3. Frequency

a) When new methods are developed in-house
b) When standard methods are applied in-house
c) When methods are changed

4. Definitions

The following definitions are derived from the USP [20] and from the ICH [18].

4.1. Method validation
The process by which it is established, by laboratory studies, that the performance characteristics of the method meet the requirements for the intended applications.

Company	Title: Validation of Analytical Methods		Code LHEQ12	Revision Number A.01
Division	Laboratory	Author	Page 2 of 9	

4.2. Specificity
The ability to assess unequivocally the analyte in the presence of components that may be expected to be present. Typically, these might include impurities, degradants, the matrix, and so on.

4.3. Accuracy
The closeness of agreement between the value that is accepted either as a conventional true value or an accepted reference value and the value found.

4.4. Precision
The closeness of agreement (degree of scatter) between a series of measurements obtained from multiple sampling of the same homogeneous sample under the prescribed conditions. Precision may be considered at three levels: repeatability, intermediate precision and reproducibility.

4.5. Repeatability
The precision under the same operating conditions over a short interval of time. Repeatability is also known as intra-assay precision.

4.6. Intermediate precision
Within-laboratory variations (e.g., different days).

4.7. Reproducibility
The precision between laboratories (collaborative studies usually applied to standardization of methodology).

4.8. Limit of detection
The lowest amount of analyte in a sample that can be detected but not necessarily quantitated as an exact value.

4.9. Limit of quantitation
The lowest amount of analyte in a sample that can be quantitatively determined with suitable precision and accuracy. The quantitation limit is a parameter of quantitative assays for low levels of compounds in sample matrices, and is used particularly for the determination of impurities and/or degradation products.

Company	Title: Validation of Analytical Methods	Code LHEQ12	Revision Number A.01
Division	Laboratory	Author	Page 3 of 9

4.10. Linearity

The ability (within a given range) to obtain test results that are directly proportional to the concentration (amount) of analyte in the sample.

4.11. Range

The interval between the upper and lower concentration (amounts) of analyte in the sample (including these concentrations) for which it has been demonstrated that the analytical procedure has a suitable level of precision, accuracy and linearity.

4.12. Robustness

The measure of a procedure's capacity to remain unaffected by small, but deliberate variations in method parameters. It provides an indication of its reliability during normal usage.

5. Determination of the Scope, Objectives and Required Performance Characteristics

5.1. Specify the objective of the method.

5.2. Specify the scope of the method.

5.3. Determine the performance characteristics and acceptance limits, for example, as shown in the following table.

Company	Title: Validation of Analytical Methods		Code	Revision Number
			LHEQ12	A.01
Division	Laboratory	Author	Page 4 of 9	

Parameter	Acceptance Limit
Specificity	

Accuracy
    amount 1
    amount 2
    amount 3	

Repeatability
    amount 1
    amount 2
    amount 3	

Intermediate precision
    amount 1
    amount 2
    amount 3	

Reproducibility	
Limit of detection	
Limit of quantitation (at a standard deviation of x%)	
Linearity	
Range	

6. Preparation for the Experiments

6.1. Materials

Specify chemicals required for the experiments, their purity and the source; allow for grade or source equivalency where applicable.

Company	Title: Validation of Analytical Methods	Code LHEQ12	Revision Number A.01
Division	Laboratory	Author	Page 5 of 9

6.2. Specify the equipment as required for the experiments; for example, Liquid Chromatograph HP1100 Series from Hewlett-Packard, with peltier-cooled thermostatted autosampler, at least binary gradient, variable wavelength detector and thermostatted column compartment. Equipment must have passed OQ test as described in SOP 453.

7. Determination of Method Performance Characteristics

7.1. Linearity and range
a) Prepare 5 standard solutions A to E containing the full working concentrations, for example, at the limit of quantitation, the upper target concentration limit A, at two concentrations between the limit of quantitation and A, and at $1.5 \times A$.
b) Inject each standard 3 times and measure the signal. Depending on the analysis technique, this may be a signal height, a peak height or a digital value as read from a display.
c) Average the results.
d) Plot the signal (area) versus the concentration. Calculate the linear regression.
e) Calculate the response factors.
Response Factor = K = signal/amount.
Compare the linear range with the specifications as set in 5.3.

7.2. Accuracy and recovery

Obtain the true value of a sample by one of these procedures:
a) Purchase certified reference materials with known amounts and uncertainty.
b) Use a reference method with known uncertainty.
c) Spike a blank sample with the analyte.

Company	Title: Validation of Analytical Methods	Code LHEQ12	Revision Number A.01
Division	Laboratory	Author	Page 6 of 9

Analyze a sample with the known true amounts at 3 different concentrations:
a) Close to the limit of quantitation
b) In the middle range
c) Close to the upper range

The sample should be processed through the entire analytical procedure, including sample preparation.

Calculate the deviation of the result obtained with the method to be validated with the true value. Compare the deviation with the criteria specified in 5.3.

7.3. Precision of amounts (repeatability)
Inject solutions A, C, and E 6 times. Calculate the standard deviation of the measured amounts. Compare the results with those specified in 5.3.

7.4. Intermediate precision
Inject solutions A, C and E on 15 working days. The analysis should be conducted by 3 different operators and 3 different columns should be used. Calculate the standard deviation of the measured amounts. Compare the results with those specified in 5.3.

7.5. Limit of detection
Prepare a standard solution with a concentration that is expected to be close to the detection limit. Inject the sample

$$\text{limit of detection (ng/L)} = \frac{3 \times \text{signal height} \times \text{standard amount (ng/L)}}{\text{baseline noise}}$$

3 times. Measure baseline noise, signal height and determine signal/noise for each measurement. Use this formula:

Average the results and compare with those specified in 5.3.

Company	Title: Validation of Analytical Methods	Code	Revision Number
		LHEQ12	A.01
Division	Laboratory	Author	Page 7 of 9

7.6. Limit of quantitation

Prepare 6 standard solutions with the amounts in the range from the expected limit of quantitation to 20 times this amount. Inject all samples 6 times and calculate the standard deviations of the amounts. Plot the standard deviations versus the amount. Take the standard deviation as specified in 5.3 and take the corresponding amount from the plot. Check if the limit of quantitation meets the criteria specified in 5.3.

8. Validation of Standard Methods

8.1. Check the scope and performance criteria of the standard method.

8.2. Check if the scope and performance criteria are within the scope of your analysis.

8.3. Check the method's text if there is any evidence that the standard methods have been validated.

8.4. Check if the results from the standard match the requirements of your method.

8.5. If either the scope of the standard method or the results are different from your criteria or if there are no validation data available, the missing validation data should be generated.

Company	Title: Validation of Analytical Methods	Code	Revision Number
		LHEQ12	A.01
Division	Laboratory	Author	Page 8 of 9

Parameter	Standard Method	User's Method	Comment
Specificity			
Accuracy			
Repeatability			
Intermediate precision			
Reproducibility			
Limit of detection			
Limit of quantitation (at a standard deviation of x%)			
Linearity			
Range			

9. Revalidation After Changes

The method should be revalidated after changing any of the following method parameters:

Change	Parameter to be validated
Concentration	Linearity, accuracy, recovery, precision
Delay volume of the HPLC pump	Selectivity (chromatographic separation)

Company	Title: Validation of Analytical Methods	Code LHEQ12	Revision Number A.01
Division	Laboratory	Author	Page 9 of 9

10. Validation report

The validation report should include the following:

10.1. Summary
A summary of the objective, the scope and a statement that the results comply with the methods intended use as specified in Section 4.

10.2. Authors of the method
The name(s), addresses and phone/fax numbers of the authors of the method.

10.3. The principle of the methods

10.4. Standards, reagents, materials and equipment used for the experiments

10.5. Procedure

10.6. Calculations

10.7. Validation results

10.8. Critical points, if any

10.9. Other comments, if any

10.10. Names and signatures of people who validated and approved the method

Example #3: Testing Precision of Peak Retention Times and Areas of an HPLC System

The following is an example of an operating procedure for the testing of an HP 1050 Series HPLC system for the precision of peak areas and retention times. This is a proposal and starting point only and may need adaptation to different HPLC systems. There is no assurance expressed that the operating procedure will pass a regulatory inspection.

Appendix C. Selected (Standard) Operating Procedures

Company	Title: Testing Precision of Peak Areas and Retention Times of an HPLC System	Code LHEQ01	Revision Number A.01
Division	Laboratory	Manager	Page 1 of 5
Effective Date	Prepared By Name: Sig: Date:	Approved By Name: Sig: Date:	Distribution — — — — —

1. Scope

Testing the precision of peak areas and retention times of an HP 1050 Series HPLC system.

2. Purpose

The precision of peak areas and retention times are important characteristics for qualitative and quantitative measurements in HPLC. This operating procedure provides chromatographic conditions and key sequences to verify these characteristics of a complete HPLC system, comprising an HP 1050 series autosampler, a quaternary pump and a variable wavelength detector.

3. Frequency

The precision should be verified at least once per year or after the repair of one or more modules.

4. Instrumentation

a) Quaternary HP 1050 series pump
b) HP 1050 series autosampler
c) HP 1050 variable wavelength detector
d) HP 3396 integrator

5. Columns, Chemicals

a) Column: 100 mm × 4.6 mm Hypersil ODS, 5 µm (HP P/N 79916OD-554)

Company	Title: Testing Precision of Peak areas and Retention Times of an HPLC System	Code LHEQ01	Revision number A.01
Division	Laboratory	Author	Page 2 of 5

b) Solvents: water and methanol, HPLC grade
c) Sample: Isocratic standard sample (Hewlett-Packard part number 01080-68704) that contains 0.15 wt.% dimethylphthalate, 0.15 wt.% diethylphthalate, 0.03 wt.% biphenyl, 0.03 wt.% o-terphenyl dissolved in methanol

6. Preparation of the Variable Wavelength Detector

a) Switch lamp ON.
b) Set the wavelength to 254 nm.
c) Set the response time to 1 SEC.

7. Preparation of the Pump

a) Prime the pump (use appropriate 1050 SOP): "Priming a Quaternary Pump."
b) Fill solvent reservoirs: A with water, B with water, C with methanol.
c) Degas solvents (see appropriate 1050 SOP): "Priming a Quaternary Pump."
d) Set UPPER LIMIT to 400 (bar).
e) Set the FLOW rate to 3.00 mL/min.
f) Set the temperature of the column oven to 45°C.
g) Set the solvent composition: A = off, B = 15%, C = 70% (channel A will be changed automatically according to %B and %C settings.
h) Set the STOP TIME to 5.00 min.
i) Switch pump ON.

8. Preparation of the Autosampler

a) Make sure that the air pressure needed for the solenoid valves is about 5 bar.
b) Switch the autosampler on.
c) Put sample vial with isocratic sample into the vial tray, position number 10.

Company	Title: Testing Precision of Peak Areas and Retention Times of an HPLC System	Code LHEQ01	Revision number A.01
Division	Laboratory	Author	Page 3 of 5

d) Set up vial numbers: FIRST 10 LAST 10.
e) Set the number of injections/vial to 6.
f) Set the injection volume to 10 µL.

9. Set Parameters for the HP 3396 Integrator

a) Attenuation: 10
b) Chart speed: 1 cm/min
c) Zero: 10
d) Threshold: 10

10. Analysis of Isocratic Standard

a) When the baseline is stable, start the analyses.
b) As a result, six chromatograms similar to the figure below should be obtained (differences in retention times and areas due to variations between different column batches and to variations in the concentration of the sample from batch to batch).

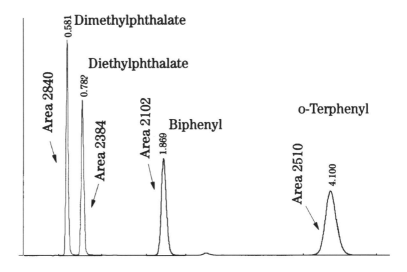

Company	Title: Testing Precision of Peak areas and Retention Times of an HPLC System	Code LHEQ01	Revision number A.01
Division	Laboratory	Author	Page 4 of 5

11. Acceptance

a) Calculate the precision of retention times and peak areas.

$$\text{RSD} = \frac{\sqrt{\frac{1}{n-1}\sum(x-\bar{x})^2}}{\bar{x}}$$

where n is the number of injections and x is the area or retention time of the peak.

$$\text{Mean} = \bar{x} = \frac{1}{n}\sum x$$

b) The precision for the peak areas should be < 1.5% RSD.
c) The precision for retention times should be < 0.5% RSD.

12. Further Action

If the 1050 HPLC system does not fulfill the given specification, do the following:

a) Check the performance of the detector (noise and drift) using appropriate SOP: "Checking Noise and Drift on the HP 1050 Series Variable Wavelength Detector."
b) Check whether the pump is leak-tight using appropriate SOP: "Leak Test for the Quaternary HP 1050 Series Pump."
c) Check whether the autosampler is leak-tight using appropriate SOP: "Checking the Pressure Tightness on HP 1050 Series Autosampler."

If following these procedures does not result in an improvement, call Hewlett-Packard service.

Company	Title: Testing Precision of Peak areas and Retention Times of an HPLC System	Code LHEQ01	Revision number A.01
Division	Laboratory	Author	Page 5 of 5

13. Appendix: Protocol Example for Results

Instrument identification
Serial number pump: _____
Serial number autosampler: _____
Serial number detector: _____
Serial number integrator: _____

Date: _____

Results
Precision of peak areas: _____ (spec < 1.5% RSD)
Precision of retention times: _____ (spec < 0.5% RSD)

Comment:

Further actions (in case the equipment is out of specification)

Approvals

	Name	Signature	Date
Laboratory supervisor	_____	_____	_____
Test engineer	_____	_____	_____

Example #4: Retrospective Evaluation and Validation of Existing Computerized Analytical Systems

The following is an example of an operating procedure for the retrospective evaluation and validation of existing computerized analytical systems. This is a proposal and starting point only and may need adaptation to different HPLC systems. There is no assurance expressed that the operating procedure will pass a regulatory inspection.

Company	Title: Retrospective Evaluation and Validation of Computerized Analytical Systems	Code LHSW03	Revision Number A.01
Division	Laboratory	Manager	Page 1 of 6
Effective Date	Prepared By Name: Sig: Date:	Approved By Name: Sig: Date:	Distribution — — — — —

1. Scope

Evaluation of existing computerized analytical systems retrospectively for past and current use and prospective validation for future use. The procedure is limited to systems purchased from a vendor.

2. Purpose

Regulatory agencies require computerized analytical systems used for the analysis and evaluation of critical data to be validated. Existing computerized systems in laboratories frequently have not been formally validated or their initial validation was not documented. The purpose of this operating procedure is to demonstrate whether such systems were operating as intended in the past, whether they are currently operating as intended and whether they will operate as intended in the future.

3. Develop a Validation Plan

Define validation requirements. Define current system expectations, evaluate what was done in the past and what is planned for the future to meet these expectations. For content and details of the plan, follow Steps 4 to 7 of this operating procedure.

Company	Title: Retrospective Evaluation and Validation of Computerized Analytical Systems		Code LHSW03	Revision Number A.01
Division	Laboratory	Manager	Page 2 of 6	

4. Describe and Define the System

a) Describe the purpose of the system.
b) List the equipment hardware:
   — in-house identification number
   — merchandising number or name
   — manufacture's name, address and phone number
   — hardware serial number, firmware revision number
   — date received in the laboratory
   — date placed in service
   — location
c) List all computer hardware:
   — manufacturer's name
   — model, serial number
   — processor, coprocessor
   — memory (RAM)
   — graphics adapter
   — hard disk
   — interfaces
   — network
d) List all software loaded on the computer software with product number, version number and the name of the vendor:
   — operating system, user interface
   — canned standard software
   — user specific application software, e.g., Macros, with date and size
e) List accessories such as cables, spare parts, etc.
f) Find and review or develop system drawings.
g) Define operator requirements.
h) Define all required functions and operational limits of the modules and system as used for the current application:

Company	Title: Retrospective Evaluation and Validation of Computerized Analytical Systems	Code LHSW03	Revision Number A.01
Division	Laboratory	Manager	Page 3 of 6

    — equipment hardware
    — software and for system functions
i) Define physical and logical security requirements, e.g., physical or password access.

5. Collect Any Documentation Available

a) Reports from internal users on number and type of problems
b) Reports from external users on number and type of problems
c) Purchase orders
d) Certificates and specifications from the vendor
e) Information on what formulas are used for calculations
f Operating procedures, for example, for basic operation, maintenance, calibration and testing of the system
g) User manuals

6. Collect Information on System History

a) Installation reports
b) Information on acceptance testing
c) System failure reports
d) Equipment hardware and system maintenance logs
e) Maintenance records
f) Calibration records
g) Results of module and system performance checks
h) Any test reports
i) Records on operator qualifications

Company	Title: Retrospective Evaluation and Validation of Computerized Analytical Systems	Code LHSW03	Revision Number A.01
Division	Laboratory	Manager	Page 4 of 6

7. Evaluate Past and Current System Performance and Document Results

Evaluate information and documentation collected under Steps 5 and 6.

a) Check to see if documentation as collected under 5f and 5g is complete and up to date. For example, does the revision of the existing user manual comply with firmware and software revision numbers?

b) Check to see if there is evidence of software development validation. Qualification criteria are availability of type and number of documents listed under 5d.

c) Check to see if the equipment (hardware) has been qualified for proper and up-to-date functions over the anticipated operating ranges as specified in 4h. Generate a matrix with equipment functions as defined in 4h versus results of calibrations and performance checks as defined in 4h.

d) Check to see if the computer system has been qualified for proper and up-to-date functions over the anticipated operating ranges as specified in 4h. Generate a matrix with system functions as defined in 4h versus results of acceptance testing. Check if calculations made by the computer software have been verified.

e) Check to see if the computerized system is suitable for its intended use as specified in 4h. Generate a matrix with performance requirements as defined in 4h versus results of system tests.

f) Check to see if the system is secure enough to meet the security requirement specifications as specified in 4i. Check also if the security features have been verified sufficiently.

Company	Title: Retrospective Evaluation and Validation of Computerized Analytical Systems	Code LHSW03	Revision Number A.01
Division	Laboratory	Manager	Page 5 of 6

g) Check to see if the number and type of errors reported under 6c indicate continuous functioning of the system.
h) Check to see if the operators were/are qualified for their jobs.
i) Prepare an evaluation report. Make a statement on past and current validation status, whether the system is formally validated (if not, define what changes to the system are needed), and make proposals for further validation steps for future use of the system.

8. Prospective Validation for Future Use

a) Update or develop system description, user requirement specifications, operating ranges, user manuals, appropriate SOPs and safety procedures as necessary.
b) Update or develop and implement a test and verification plan for the equipment. The plan should be developed to verify the performance of the various equipment parameters over the anticipated operating ranges and should include documented test procedures, expected results and acceptance criteria. After the test phase a formal report that documents the results should be generated.
c) Update, develop and implement an acceptance test plan for the computer system. Develop a test plan to exercise the various functions of the computer system. Specify the functions to be tested, the purpose of the individual tests, the test steps or methodology, the expected results and the acceptance criteria. Develop test cases and test data sets with known inputs and outputs for functional testing. Include test cases with normal data across the operating range, boundary testing and unusual cases (wrong inputs). After the test phase, a formal report that documents the results should be generated.

Company	Title: Retrospective Evaluation and Validation of Computerized Analytical Systems		Code LHSW03	Revision Number A.01
Division	Laboratory	Manager	Page 6 of 6	

d) Update or develop and implement an operator qualification plan.
e) Update or develop and implement a preventive maintenance plan.
f) Update or develop and implement a calibration schedule and/or a performance verification schedule.
g) Update or develop and implement a procedure for annual system review.
h) Update or develop and implement an error recording, reporting and remedial action plan.

9. Approvals

The validation plan, the system definition, the results of past and current evaluation, the prospective validation plan and test plans and results should be approved and signed by the user and the QA departments.

10. References

a) N. R. Kuzel, Fundamentals of computer system validation and documentation in the pharmaceutical industry, *Pharm. Tech.*, Sept. 1985, pp. 60–76.
b) J. Agalloco, Validation of existing computer systems, *Pharm. Tech.*, Jan. 1987, pp. 38-40.
c) H. Hambloch, Existing computer systems: A practical approach to retrospective evaluation, *Computer validation practices*, pp. 93–112, Buffalo Grove, Ill., USA, Interpharm Press, Inc., ISBN 0-935184-5-4, 1994.
d) L. Huber, *Validation of computerized analytical systems*, Buffalo Grove, Ill., USA, Interpharm Press, Inc., ISBN 0-935184-75-9, May 1995.

Appendix D. Selected Case Studies for OQ

Scenario 1: Pharmaceutical QC Lab with 25 Existing HPLC Systems

The systems have been acquired from 3 different vendors over the last 10 years. They are used for compound analysis in the range of 20 to 100 percent. Of the 25 systems, 3 are used for compound analysis and simultaneously for quantitative analysis of impurities down to 0.1 percent. They are all equipped with an isocratic pump, with an automated sampler, and with a time programmable variable wavelength UV/visible detector. The detector's baseline noise manufacturer's specification varies from 1.5×10^{-5} AU (best) to 1×10^{-4} (worst). The instrument's wavelength accuracy varies from ±1 nm to ±4 nm. Only systems recently acquired from Hewlett-Packard are equipped with built-in holmium oxide filters for wavelength calibration.

A few remarks:

1. An SOP should be developed that will be used for all systems. The test procedures are all the same.

2. The acceptance limits are all the same. The baseline noise specification is 5 times the specification of the worst instrument. This provides enough tolerance in case the condition of the system is not in ideally matched with new maintenance parts. It also satisfies the needs of most applications run in the laboratory. Only the ones that are used for main compounds and impurities will have a more stringent acceptance limit of 5×10^{-5} AU.

Table 1. Test parameters and acceptance criteria for Scenario 1.

Parameter	Procedure (*)	User Limit
Leak testing	Flow test by volume/time	±5%
Baseline drift	ASTM Method E19.09, 20 min	2×10^{-3} AU
Baseline noise	ASTM Method E19.09, 20 × 1 min	2×10^{-4} AU (or 5×10^{-5} AU)
Precision of injection volume	6 injections of caffeine standard, RSD of peak areas	2% RSD
Precision of flow rate	6 injections of caffeine standard, RSD of retention times	2% RSD
Detector linearity	Inject 5 standards	1.5 AU, 5%
Wavelength accuracy	Certified caffeine standard is injected	±4 nm
Autosampler carryover	Injection of blank solvent after large concentration	0.3%

(*) For detailed procedure, see Reference 101.

3. Even if some units have a built-in holmium oxide filter for wavelength accuracy calibration, a certified caffeine standard is injected to verify the wavelength accuracy of all instrments.

4. At the beginning of the test, a leak test based on the instrument's flow rate accuracy is performed. If this test fails, most other tests are likely to fail.

5. The instruments used for quantitative analysis are always calibrated with a chemical standard before and during a series of sample analyses. Therefore, the accuracy and linearity of the injection volume does not need to be tested.

6. The OQ procedure should be scheduled once a year.

Scenario 2: Environmental Testing Lab Equipped with a HPLC System

The HPLC system is an HP1100 Series HPLC system with a quaternary pump, column compartment, automated sampler with 100 vials, diode-array detector (DAD) and a ChemStation. The system is intended to be used mainly for the analysis of phenyl urea herbicides in drinking water. The limit of quantitation is 50 ppt. With an injection volume of 50 µL, this results in a peak height of about 100 mAU. The DAD has

Table 2. Test parameters and acceptance criteria for Scenario 2.

Parameter	Procedure (*)	User Limit
Leak testing	Flow test by volume/time	±5%
Baseline drift	ASTM Method E19.09, 20 min	2×10^{-3} AU
Baseline noise	ASTM Method E19.09, 20 × 1 min	6×10^{-5} AU
Precision of injection volume	6 injections of caffeine standard, RSD of peak areas	1% RSD
Precision of flow rate	6 injections of caffeine standard, RSD of retention times	1% RSD
Detector linearity	Inject 5 standards	1.0 AU, 5%
Wavelength accuracy	Holmium oxide filter	±2 nm
Temperature accuracy	Comparison with external measuring device	±1°C
Temperature precision	Monitoring temperature over 20 min	±0.25°C
Autosampler carryover	Injection of blank solvent after large concentration	< 0.3%
Mobile phase composition accuracy	Step gradients from 4 to 7% B, step heights relative to 100% with acetone tracer	±1%
Mobile phase composition ripple	Peak-to-peak noise at 4, 5, 6 and 7% B	0.2%

(*) For detailed procedure, see Reference 101.

built-in holmium oxide filter for wavelength calibration. The specification for baseline noise is 2×10^{-5} AU.

A few remarks:

1. For wavelength calibration, we recommend using the built-in holmium oxide filter.

2. We recommend setting the limit for the baseline noise of the UV/visible DAD to 6×10^{-5} AU. This is 3 times higher than the instrument specification. This provides enough tolerance in case the condition of the systems is not ideally matched with new maintenance parts.

3. The OQ procedure should be scheduled once a year. Because the detector's baseline noise is critical for the success of the application, this test should be scheduled every month.

4. As long as the instrument is used for pesticides only, a pesticide standard compound, e.g., triazine, can be used for flow precision, injection precision and a linearity test, if a suitable certified standard is available.

5. The instrument is always calibrated with a chemical standard before and during a series of sample analyses. Therefore, the accuracy of the injection volume does not need to be tested.

Appendix E. Books in the Area of Qualification and Validation

Quality Assurance/Quality Control

Quality Assurance of Chemical Measurements

- Key words: quality assurance, statistics, sampling, calibration, validation
- Author: John Keenan Taylor
- Publisher: CRC Press, Inc, 1987, ISBN: 0-87371-097-5
- Comment: 328 pages; published in 1987 but still considered to be the reference book for QA of chemical measurements.

Guidelines for Laboratory Quality Auditing

- Key words: quality assurance, audits, accreditation, OECD GLP, U.S. FDA GLP
- Authors: Donald C. Singer and Ronald P. Upton
- Publisher: ASQC Quality Press, Marcel Dekker, 1993, ISBN: 0-8247-8784-6
- Comment: 411 pages with 43 pages of text and over 300 pages with appendices on EPA, FDA and the OECD GLP regulations.

Accreditation and Quality Assurance in Analytical Chemistry

- Key words: quality assurance, accreditation, statistics, sampling, method validation, traceability, reference material, accreditation versus GLP, EURACHEM, U.S. EPA

- Editor: Helmut Günzler
- Publisher: Springer-Verlag, 1996, ISBN: 3-540-60103-1
- Comment: 265 pages; multiauthor book with very comprehensive and detailed information on the topics listed under key words.

Guide to Quality Control

- Key words: control charts, pareto diagrams
- Author: Kaoru Ishikawa
- Publisher: Nordica International Limited, 1990, ISBN 92-833-1035-7
- Comment: 225 pages; detailed information on how to collect data and how to construct and interpret QC charts, pareto diagrams, scatter diagrams and so on.

Quality Assurance for Analytical Laboratories

- Key words: quality assurance, accreditation, statistics, GLP, reference material
- Author: M. Parkany
- Publisher: Royal Society of Chemistry, 1993, ISBN: 0-85186-705-7
- Comment: 197 pages; includes excellent chapter on people motivation.

Quality Assurance and TQM for Analytical Laboratories

- Key words: quality assurance, accreditation, statistics, quality audits, reference material, method validation
- Editor: M. Parkany
- Publisher: Royal Society of Chemistry, 1995, ISBN: 0-85404-760-3

Comment: 287 pages

Quality in the Analytical Chemistry Laboratory

- Key words: quality assurance, sampling, method selection, equipment, reference material, statistics, reporting, uncertainty, costs
- Coordinating Author: E. Prichard

- Other Authors: Neil T. Crosby, John A. Day, William A. Hardcastle, David G. Holcombe and Ric D. Treble
- Publisher: John Wiley & Sons, 1995, ISBN: 0-471-95470-5
- Comment: 307 pages; very practical book with many checklists that help to implement a quality system.

Quality Assurance Principles for Analytical Laboratories

- Key words: quality assurance, statistics, control charts, people qualification, equipment, sample handling, sampling, control samples, audits, safety, accreditation, quality manual
- Author: Frederick M. Garfield
- Publisher: AOAC International, 1991, ISBN: 0-935584-46-3
- Comment: Comprehensive information on all aspects of QA in a chemical laboratory. Excellent chapter on using QC charts. Good summary tables of equipment qualification.

Quality Control in Analytical Chemistry

- Key words: quality assurance, sampling, data processing, costs
- Authors: G. Kateman and L. Buydens
- Publisher: John Wiley & Sons, Inc, 1993, ISBN: 0-471-55777-3
- Comment: second edition with 317 pages; includes large chapters on sample handling, sampling and data handling.

Laboratory Accreditation and Data Certification

- Key words: accreditation, certification, data, U.S. EPA
- Authors: Carla H. Dempsy and J. D. Petty
- Publisher: Lewis Publisher, 1991, ISBN: 0-87371-291-9
- Comment: 240 pages; presents a system for laboratory accreditation in conjunction with data certification that assures purchasers usefulness of data.

Writing the Laboratory Notebook

- Key words: notebook, GLP, GMP
- Author: Howard M. Kanare
- Publisher: American Chemical Society, 1985, ISBN: 0-8412-0906-5
- Comment: 145 pages; detailed information in writing a laboratory paper notebook.

Computer Validation

Validation of Computerized Analytical Systems

- Key words: validation, computers, international regulations, pharmaceutical industry
- Author: Ludwig Huber
- Publisher: Interpharm Press, Inc., 1995, ISBN: 0-935184-75-9
- Comment: 267 pages. Readers will learn what is required for the validation of computerized analytical systems and how to do it right the first time. Includes many checklists and SOPs for easy implementation.

Validating Automated Manufacturing and Laboratory Applications

- Key words: validation, computers, international regulations, pharmaceutical industry, manufacturing
- Editor: Guy Wingate
- Publisher: Interpharm Press, Inc., 1997, ISBN: 0-57491-037-X
- Comment: 564 pages; includes 15 practical case studies.

Computer Validation Compliance: A Quality Assurance Perspective

- Key words: validation, computer systems
- Authors: Mary Ellen Double and Maryann McKendry
- Publisher: Interpharm Press, Inc., 1994, ISBN: 0-935184-48-1

- Comment: Binder format, 205 pages; fundamentals of computer validation; includes checklists for the inspection of computer systems and SOPs for validation. The book has been written by a senior vice president of Merck Research Laboratories and presents the key elements of Merck's QA philosophy and practices.

Good Computer Validation Practices: Common Sense Implementation

- Key words: validation, computers, SOPs, international regulations, life cycle, pharmaceutical industry
- Authors: Teri Stokes, Ronald C. Branning, Kenneth G. Chapman, Heinrich Hambloch and Anthony J. Trill
- Publisher: Interpharm Press, Inc., 1994, ISBN: 0-935184-55-4
- Comment: 324 pages; written by a group of international experts on computer validation and covers aspects of new and existing systems.

ISO 9000-3: A Tool for Software Product and Process Improvement

- Key words: software, validation, computers, life cycle
- Authors: Raymond Kehoe and Alka Jarvis
- Publisher: Springer, 1995, ISBN: 0-38794568-7
- Comment: 229 pages; good interpretation and further guidelines to implement the ISO 9000-3 quality guide for software development, marketing and support.

Computer Systems Validation for the Pharmaceutical and Medical Device Industries

- Key words: validation, computer, IQ, OQ, product life cycle
- Author: Richard Chamberlain
- Publisher: Alaren Press, 1991, ISBN: 0-9631489-0-7
- Comment: 217 pages with fundamentals of computer validation and 60 pages of draft SOPs. The draft SOPs are included on a floppy disk in WordPerfect and ASCII format.

GLP/GMP computer procedures and systems

- Key words: validation, computer, GLP, GMP
- Author: Donald L. Crone
- Publisher: Interpharm Press, Inc., 1995, ISBN 0-935184-74-0
- Comment: binder format, > 400 pages, useful forms and SOPs for easy implementation. The draft SOPs are included on a floppy disk in WordPerfect and ASCII format.

Standards, Guidelines, and Examples on System and Software Requirements Engineering

- Key words: computer, software, validation, life cycle
- Author: Merlin Dorfman and Richard H. Thayer
- Publisher: IEEE Computer Society Press, 1990, ISBN: 0-8186-8922-6
- Comment: 605 pages; very detailed guidelines on good software engineering practices, with a focus on requirement specifications. Includes about 70 pages of glossary terms related to the topic.

Method Validation/Statistics

Use of Statistics to Develop and Evaluate Analytical Methods

- Key words: statistics, method validation, interlaboratory studies, AOAC
- Author: Grant T. Wernimont
- Publisher: Association of Official Analytical Chemists, 1985, ISBN: 0-935584-31-5
- Comment: 183 pages; focuses on interlaboratory evaluation of analytical procedures; contains detailed statistics.

Development and Validation of Analytical Methods

- Key words: regulations, pharmaceutical, robotics, out-of-specification situations, dissolution studies, biological samples
- Editors: Christopher M. Riley and Thomas W. Rosanske

- Publisher: Pergamon Press, 1996, ISBN: 0-08042792-8
- Comment: 352 pages. The book includes an up-to-date review on international regulations and guidelines related to the topic. The chapter on interpretation of the cGMP issues contained in the *United States v. Barr Laboratories* case is extremely valuable for pharmaceutical QA/QC laboratories. Also, the chapter on validation of methods for biological samples is very useful and provides detailed instructions related to the topic.

Analytical Method Development and Validation

- Key words: method validation, pharmaceutical, ICH, FDA
- Authors: Michael E. Swartz and Ira Krull
- Publisher: Marcel Dekker, 1997, ISBN: 0-0247-0115-1
- Comment: 92 pages

ISO 9000

ISO 9000: Preparing for Registration

- Key words: ISO 9000 series, quality pyramid, documentation, costs, audits, quality manual
- Author: James L. Lamprecht
- Publisher: ASQC Quality Press, Marcel Dekker, 1992, ISBN: 0-8247-8741-2
- Comment: 241 pages; useful appendix with guideline on what to include in a quality manual.

Implementing the ISO 9000 Series

- Key words: ISO 9000 series, implementation, change control, internal audits, future of ISO 9000
- Author: James L. Lamprecht
- Publisher: ASQC Quality Press, Marcel Dekker, 1993, ISBN: 0-8247-9134-7
- Comment: 262 pages; focuses on how to implement the ISO 9000 series.

Food Laboratory

Food and Drink Laboratory Accreditation

- Key words: quality assurance, accreditation, NAMAS, audits
- Authors: Sandra Wilson and Geoff Weir
- Publisher: Chapmann & Hall, 1995, ISBN: 0-412-59920-1
- Comment: 262 pages; useful book on how to develop, obtain and maintain accreditation; examples of a NAMAS quality manual help with implementation.

Regulations and Guidelines: GLP, GALP, cGMP, GCP, . . .

Implementing International Good Practices

- Key words: regulations, international, GMP, GLP, GCP, OECD
- Author: Nigel J. Dent
- Publisher: Interpharm Press, Inc., 1993, ISBN: 0-935184-44-9
- Comment: 284 pages. Very good overview about worldwide GLP and GCP regulations. Focus on OECD GLPs, GMP in Scandinavia, GLP in Japan, GCPs in United States and Europe and approaches to implementation.

FDA–SPEAK: The Interpharm Glossary of Acronyms and Regulatory Terms

- Key words: regulations, GMP, GLP, glossary
- Author: Dean E. Snyder
- Publisher: Interpharm Press, Inc., 1992, ISBN: 0-935184-30-9
- Comment: 267 pages. A common language on aspects of good practices and validation is very important for a good understanding of the topics. This book provides the terminology of the U.S. FDA and is recommended for those who have involvement with the FDA. Also includes organization charts, addresses and phone numbers of the FDA.

Good Laboratory Practice and Current Good Manufacturing Practice

- Key words: GLP, cGMP, HPLC, GC, MS, CE, UV/vis
- Author: Ludwig Huber
- Publisher: Hewlett-Packard, 1994, Publication number: 12-5963-2115E; has been translated into 10 different languages.
- Comment: 152 pages; provides a summary of the basics of GLP and cGMP and its impact on analyses with HPLC, GC, MS, CE and UV/vis spectroscopy.

GCP Quality Audit Manual

- Key words: GCP, audits
- Author: James E. Sayre
- Publisher: Interpharm Press, Inc., 1990, ISBN 0-935184-56-2
- Comment: binder format, 60 pages of text, 56 pages of audit checklists, 173 pages of regulatory text

GALP Regulatory Handbook

- Key words: GALP, validation, computers, regulations
- Authors: Weinberg, Spelton and Sax
- Publisher: Lewis Publishers, 1994, ISBN: 1-56670-025-6
- Comment: ~ 230 pages; assists laboratories in applying the U.S. EPA's GALPs from 1990 (does not include changes made to the final recommendations in 1995). The book includes the EPA's text from 1990 and comments the individual chapters. Weinberg, Spelton and Sax, Inc., is a privately held consulting organization.

International GLPs

- Key words: GLP, regulations, international, memorandum of understandings
- Authors: Robert S. DeWoskin and Stefanie M. Taulbee
- Publisher: Interpharm Press, Inc., 1993, ISBN: 0-935184-42-2

- Comment: 452 pages, binder format; side-by-side comparison of the most well-known national and international GLP standards. Includes the full text of most GLPs: U.S. EPA; U.S. FDA; OECD; Japanese Ministry of Health and Welfare (MHW); Japanese Ministry of Agriculture, Forestry and Fisheries; and Japanese Ministry of International Trade and Industry. It also includes the text of many memoranda of understandings (MOU).

International GMPs

- Key words: GMP, regulations, international
- Authors: Michael H. Anisfeld
- Publisher: Interpharm Press, Inc., 1993, ISBN: 0-935184-17-1
- Comment: more than 500 pages, binder format; summarizes the worldwide GMP requirements and countries performing international GMP inspections. The texts of international conventions and requirements (WHO, Asian, EU, PIC–GMP, PIC–Bulk), plus the text of more than 20 national GMPs are included. Also lists national regulatory agencies responsible for drug manufacturing regulations.

International Biotechnology, Bulk Chemical, and Pharmaceutical GMPs, 5th ed.

- Key words: GMP, regulations, international
- Author: Michael H. Anisfield
- Publisher: Interpharm Press, Buffalo Grove, IL, 1998, ISBN: 1-57491-043-4
- Comment: more than 500 pages, binder format. Summary of worldwide Good Manufacturing Requirements. Summary of countries performing international GMP inspections. Texts of international conventions and requirements (WHO, Asian, EU, PIC–GMP, PIC–Bulk). Text of more than 20 national GMPs. Listing of national regulatory agencies responsible for drug manufacturing regulations.

Validation Compliance Annual 1995

- Key words: regulations, validation, electronic signatures, inspections, pharmaceutical, computer validation

- Author: International Validation Forum
- Publisher: ASQC Quality Press, Marcel Dekker, ISBN: 0-8247-9459-1
- Comment: 1,100 pages. Detailed U.S. FDA, U.S. EPA and EU validation requirements and techniques for the validation processes. The International Validation Forum is a not-for-profit association offering training and certification of validation professionals.

Good Laboratory Practice Regulations

- Key words: GLP, U.S. FDA, international regulations
- Author: Allen F. Hirsch
- Publisher: ASQC Quality Press, Marcel Dekker, 1989, ISBN: 0-8247-8101-5
- Comment: 234 pages. Though published in 1989, it is still the reference book on the basics of GLP. It includes many practical examples for implementation.

Good Laboratory Practice Standards

- Key words: GLP, SOPs, field studies, computer validation, U.S. EPA
- Authors: Willa Y. Garner, Maureen S. Barge and James P. Ussary
- Publisher: American Chemical Society, 1992, ISBN: 0-8412-2192-8
- Comment: 571 pages; very detailed book on implementing GLPs. Focuses on the EPA GLPs. It also includes appendices with the text of GLP standards, standard forms for submitting data to the U.S. EPA and a question/answer section.

Pharmaceutical

Quality Assurance for Biopharmaceuticals

- Key words: quality assurance, regulations, GMP, biopharmaceutical, sampling, method validation
- Author: Jean F. Huxsoll

- Publisher: John Wiley & Sons, 1994, ISBN: 0-471-03656-0
- Comment: 206 pages; tailored to the biopharmaceutical manufacturing process.

Qualification and Validation in Pharmaceutical Manufacture

- Key words: quality assurance, regulations, GMP, biopharmaceutical, sampling, method validation
- Authors: Seminar speakers from a seminar held in Dublin
- Publisher: Published by the Secretariat to the Convention for the Mutual Recognition of Inspections in Respect of the Manufacture of Pharmaceutical Products (EFTA Secretariat, 9-11 rue de Varembe, CH-1211 Geneva 20), 1994
- Comment: 287 pages; includes useful information on regulatory requirements and solutions as practiced in the pharmaceutical industry. Some chapters are very detailed. Good chapter on "How much validation is enough?."

Documentation Basics That Support Good Manufacturing Practices

- Key words: GMP, documentation
- Author: Carol DeSain
- Publisher: Aster Publishing Corporation, 1993, ISBN: 0-943330-30-0
- Comment: 88 pages. Describes documentation requirements for cGMP and gives recommendations on who writes what. Includes many forms for easy implementation.

Training for the Healthcare Manufacturing Industries: Tools and Techniques to Improve Performance

- Key words: pharmaceutical, people, training
- Authors: James L. Vesper
- Publisher: Interpharm Press, Inc., 1993, ISBN: 0-935184-43-0
- Comment: 413 pages; examines topics related to training, adult learning, human performance and new training technologies. Includes evaluation forms.

Appendix F. References

1. U.S. FDA, Title 21 of the U.S. Code of Federal Regulations: 21 CFR 211—*Current good manufacturing practice for finished pharmaceuticals.*

2. U.S. FDA, *Technical review guide: Validation of chromatographic methods*, Center for Drug Evaluation and Research (CDER), Rockville, Md., USA (1993).

3. U.S. EPA, *Guidance for methods development and methods validation for the Resource Conservation and Recovery Act (RCRA) Program*, Washington, D.C. (1995).

4. AOAC *Peer-Verified Methods Program*, Manual on policies and procedures, Arlington, Va., USA (November 1993).

5. ISO 7870:1993, Control charts—general guide and introduction.

6. ISO 7873:1993, Control charts for arithmetic average with warning limits.

7. ISO 7966:1993, Acceptance control charts.

8. ISO 8258:1991, Shewhart control charts.

9. ISO/IEC Guide 32: *Calibration of chemical analysis and use of certified reference materials* (in preparation).

10. ISO/IEC Guide 33: *Use of certified reference material* (1989).

11. ISO/IEC Guide 35: *Certification of reference materials—general and statistical principles* (1989).

12. PMA, *Concepts and principles for the validation of computer systems in the pharmaceutical industry* (includes 33 papers), Proceedings of CSVC seminar in Crystal City, Va., USA, 15–18 January 1984.

13. PMA, *Concepts and principles for the validation of computer systems used in the manufacture and control of drug products* (includes 24 papers on contemporary computer validation subjects), Proceedings of CSVC Seminar II in Chicago, 20–23 April 1986.

14. M. Freeman, M. Leng, D. Morrison and R. P. Munden from the UK Pharmaceutical Analytical Sciences Group (PASG), Position Paper on the qualification of analytical equipment, *Pharm. Techn. Europe*, November 1995, 40–46.

15. P. Bedson and M. Sargent, The development and application of guidance on equipment qualification of analytical instruments, *Accred. Qual. Assur.* 1 (6):265–274 (1996).

16. J. Sharp, *Validation—how much validation is required*, Proceedings: Convention of the mutual recognition of inspections in respect of the manufacture, Qualification and validation in pharmaceutical manufacture, Dublin, 1994, p. 266.

17. EPA, *Good automated laboratory practices, principles and guidance to regulations for ensuring data integrity in automated laboratory pperations, with implementation guide*, Research Triangle Park, N.C., USA (1995).

18. International Conference on Harmonization (ICH) of Technical Requirements for the Registration of Pharmaceuticals for Human Use, *Validation of analytical procedures: definitions and terminology*, Geneva (1996).

19. International Conference on Harmonization (ICH) of Technical Requirements for the Registration of Pharmaceuticals for Human Use, *Validation of analytical procedures: Methodology*, adopted in 1996, Geneva.

20. General Chapter 1225, Validation of compendial methods, *United States Pharmacopeia* 23, National Formulary 18, Rockville, Md., USA, The United States Pharmacopeial Convention, Inc., pp. 1982–1984 (1995).

21. EURACHEM Guidance Document No. 1/WELAC Guidance Document No. WGD 2: *Accreditation for chemical laboratories: Guidance on the interpretation of the EN 45000 series of standards and ISO/IEC Guide 25* (1993). Available from the EURACHEM Secretariat, PO Box 46, Teddington, Middlesex, TW11 ONH, UK, tel.: +44 81 943 7614, fax: +44 1 81 943 2767.

22. EURACHEM, *Quantifying uncertainty in analytical measurement*, ISBN 0-948926-08-2, Teddington, UK (1995).

23. CITAC, Working Group, *International guide to quality in analytical chemistry: An aid to accreditation*, Teddington Middlesex, UK, Edition 1.0 (1995).

24. U.S. FDA GLP, *Good laboratory practice regulations for non-clinical studies*, Final rule, U.S. FDA, Rockville, Md., USA, Title 21 CFR, Part 58 (1979).

25. U.S. EPA, Federal Insecticide, Fungicide and Rodenticide Act (FIFRA): *Good laboratory practice standards, Federal Register* 48 (230):53946–53969; Nov. 29, 1983; effective May 2, 1984.

26. U.S. EPA, Toxic substance control act (TSCA): Good laboratory practice standards, 40 CFR Part 792, *Federal Register* 54 (158):34034–34052; Aug. 17, 1989; effective Sept. 18, 1989.

27. Organization of Economic Co-operation and Development, *Good laboratory practice in the testing of chemicals*, final report of the Group of Experts on Good Laboratory Practice (1982, out of print).

28. OECD, The OECD principles of good laboratory practice, *Series on principles of good laboratory practice and compliance monitoring*, number 5, GLP consensus document environment monograph No. 45, Paris (1992).

29. European Community, *The harmonization of laws, regulations and administrative provisions to the application of the principles of good laboratory practice and the verification of their application for tests on chemical substances*, Council Directive 87/18/EEC (1987).

30. European Community, *The inspection and verification of good laboratory practice*, Council Directive 88/320/EEC (1988), adopted in 1990 (90/18/EEC).

31. H. Anisfeld, *International Biotechnology, Bulk Chemical, and Pharmaceutical GMPs*, 5th ed., Buffalo Grove, Ill., USA, Interpharm Press, Inc. (1998).

32. U.S. FDA, *Guide to Inspection of pharmaceutical quality control Laboratories*, Final rule, U.S. FDA, Rockville, Md., USA (1993).

33. Commission of the European Communities, EC guide to good manufacturing practice for medicinal products, in: *The rules governing medicinal products in the European Community*, Volume IV, Office for Official Publications for the European Communities, Luxembourg, ISBN 92-826-3180-X105 (1992).

34. ICH second International Conference on Harmonization, Closing report on the Status of Tripartite Harmonization Initiatives, Orlando, Fla., USA (October 1993).

35. General Chapter 621, *United States Pharmacopeia 23*, National Formulary 18, Rockville, Md., USA, The United States Pharmacopeial Convention, Inc., pp. 1776–1777 (1995).

36. EN 45001:1989, *General criteria for the operation of testing laboratories*, Rue Brederode 2, B-1000 Brussels, CEN/CENELEC, The Joint European Standard Institution.

37. ISO/IEC Guide 25: *General requirements for the competence of calibration and testing laboratories*, 3rd ed. (1990). International Organization for Standardization, Case postale 56, CH-1211 Geneve 20, Switzerland. (As of 1998, the guide is under revision.)

38. ISO 9000-3:1991(E), Quality management and quality assurance standards, Part 3: *Guidelines for the application of ISO 9001 to the development, supply and maintenance of software*, International Organization for Standardization, Case postale 56, CH-1211 Geneve 20, Switzerland.

39. European Information Technology Quality System Auditor Guide, January 1992, document 92/0001 (ITAGC) and SQA-68 (CTS-2). Available from ITQS Secretariat, AIB-Vincotte, Avenue du Roi 157, B-1060 Brussels, Belgium; Internet: http://www.nsai.ie/itqs.htm.

40. *The TickIT guide to software quality management system construction and certification using EN 29001*, Issue 2, ISBN 0-9519309-0-7 (Feb. 1992). Available from the DISC TickIT Office, 2 Park Street, London W1A 2BS, UK, tel: +44 1 716028536, fax: +44-1-716028912, http://www.avnet.co.uk/tesseract/QiC/TickIT.

41. EURACHEM-UK Instrumentation Working Group, *Guidance on equipment qualification (EQ) of high performance liquid chromatography (HPLC) instruments*, Teddington, Middlesex, UK, Draft (1997).

42. *Voluntary Guideline for IQ, OQ and PQ,* The Analytical Instrument Association (AIA), Alexandria, Va., USA (1996).

43. U.S. FDA, *General principles of validation,* Rockville, Md., USA, Center for Drug Evaluation and Research (CDER) (May 1987).

44. World Health Organization, Good practices for the manufacture and quality control of drugs, in: *International drug GMPs,* Buffalo Grove, Ill., USA, Interpharm Press, Inc., ISBN 0-935184-17-1 (1993).

45. ASEAN GMP Guide: Good Manufacturing Practice, General Guidelines, in: *International Drug GMPs,* Buffalo Grove, Ill., USA, Interpharm Press, Inc., ISBN 0-935184-17-1 (1990).

46. Pharmaceutical Inspection Convention (PIC), Guide to Good Manufacturing Practice for Pharmaceutical Products, published in: *International drug GMPs,* Buffalo Grove, Ill., USA, Interpharm Press, Inc., ISBN 0-935184-17-1 (1993).

47. OECD, *The application of the principles of GLP to computerized systems, Series on principles of good laboratory practice and compliance monitoring,* GLP consensus document number 10, environment monograph No. 116, Paris (1995).

48. K. G. Chapman, A history of validation in the United States, Part I, *Pharm. Tech.,* Oct. 1991, pp. 82–96.

49. L. Huber, *Validation of computerized analytical systems,* Buffalo Grove, Ill., USA, Interpharm Press, Inc., ISBN 0-935184-75-9 (1995), Hewlett-Packard part number 5959-3879.

50. PMA's Computer System Validation Committee, Validation concepts for computer systems used in the manufacture of drug products, *Pharm. Technol.* 10 (5):24–34 (May 1986).

51. W. B. Furman, T. P. Layloff and R. F. Tetzlaff, Validation of computerized liquid chromatographic systems, paper presented at the Workshop on Antibiotics and Drugs in Feeds, 106th AOAC Annual International Meeting and Exposition, 30 August 1992, Ohio, USA; published in *J. AOAC Intern.* 77 (5):1314–1318 (1994).

52. V. P. Shah et al., Analytical methods validation: Bioavailability, bioequivalence and pharmacokinetic studies. *Eur. J. Drug Metabolism and Pharmacokinetics* 16 (4):249–255 (1991) **and** *Pharm. Res.* 9:588–592 (1992).

53. U.S. FDA, *Guidelines for submitting samples and analytical data for method validation*, Rockville, Md., USA, Center for Drugs and Biologics, Department of Health and Human Services (Feb. 1987).

54. U.S. FDA HSS, 21 CFR Part 211, *Current good manufacturing practice of certain requirements for finished pharmaceuticals*, Proposed Rule (May 1996).

55. G. C. Hokanson, A life cycle approach to the validation of analytical methods during pharmaceutical product development, part I: The initial validation process, *Pharm. Tech.*, Sept. 1994, pp. 118–130.

56. G. C. Hokanson, A life cycle approach to the validation of analytical methods during pharmaceutical product development, Part II: Changes and the need for additional validation, *Pharm.Tech.*, Oct. 1994, pp. 92–100.

57. J. M. Green, A practical guide to analytical method validation, *Anal. Chem. News & Features*, 1 May 1996, pp. 305A–309A.

58. B. Renger, H. Jehle, M. Fischer and W. Funk, Validation of analytical procedures in pharmaceutical analytical chemistry: HPTLC assay of theophylline in an effervescent tablet, *J. Planar Chrom.* 8:269–278 (July/August 1995).

59. Wegscheider, Validation of analytical methods, in: *Accreditation and quality assurance in analytical chemistry*, edited by H. Guenzler, Springer Verlag, Berlin (1996).

60. S. Seno, S. Ohtake and H. Kohno, Analytical validation in practice at a quality control laboratory in the Japanese pharmaceutical industry, *Accred. Qual. Assur.* 2:140–145 (1997).

61. P. A. Winslow and R. F. Meyer, Defining a master plan for the validation of analytical methods, *J. Validation Technology*, pp. 361–367 (1997).

62. G. Szepesi, M. Gazdag and K. Mihalyfi, Selection of HPLC methods in pharmaceutical analysis—III method validation, *J. Chromatogr.* 464:265–278.

63. J. Vessman, Selectivity or specificity? Validation of analytical methods from the perspective of an analytical chemist in the pharmaceutical industry, *J. Pharm & Biomed Analysis* 14:867–869 (1996).

64. L. Huber, *Applications of diode-array detection in HPLC*, Waldbronn, Germany, Hewlett-Packard, publ. number 12-5953-2330 (1989).

65. L. Huber and S. George, *Diode-array detection in high-performance liquid chromatography*, New York, Marcel Dekker, ISBN 0-8247-4 (1993).

66. D. Marr, P. Horvath, B. J. Clark, A. F. Fell, Assessment of peak homogeneity in HPLC by computer-aided photodiode-array detection, *Anal. Proceed.* 23:254–257 (1986).

67. ISO/IEC Guide 25: *General requirements for the competence of calibration and testing laboratories*, draft 4 (1996). International Organization for Standardization, Case postale 56, CH-1211 Geneve 20, Switzerland.

68. ISO/IEC Guide 30:1992: *Terms and definitions used in connection with reference materials.*

69. J. K. Taylor, *Quality assurance of chemical measurements*, Lewis Publishers, Inc., 121 South Main Street, P.O. Drawer 519, Chelsea, MI 48118, ISBN 0-87371-097-5 (1990).

70. A. Gaskill, The standard's question, *Environmental Lab.*, June/July 1991, pp. 40/41.

71. W. R. Russo, Standards certification program, *Environmental Lab.*, June/July 1991, pp. 42–43.

72. M. Bolgar, Third party certification, *Environmental Lab.*, June/July 1991, 446–447.

73. M. Kubota, K. Kato, A. Hioti, H. Iijima and M. Matsumoto, Development and supply of reference materials based on the Measurement Law in Japan, *Accred Qual Assur.* 2:130–136 (1997).

74. Z. E. Bunn, The standards dilemma, *Environmental Lab.*, June/July 1991, pp. 48–49.

75. A. Gaskill, News and views: Environmental reference standards, *Environmental Lab.*, Feb/March 1990, pp. 12–15.

76. *Directory of certified reference materials*, Secretary for REMCO, ISO, Case postale 56, 1211 Geneva, Switzerland.

77. Physicochemical Measurements: Catalogue of Reference Materials from National Laboratories, *Pure Appl. Chem.* 48:503 (1976).

78. *Certified reference materials,* catalogue of the Office of Reference Materials, LGC, Queensroad, Teddington, Middlesex TWY11 OLY, UK, p. 5 (1993).

79. J. K. Borchardt, Mastering the skills of the job interviewer, *Today's Chemists at Work,* November 1997, pp. 31–33.

80. R. E. Lawn, M. Thompson and R. F. Walker, Proficiency testing in analytical chemistry, LGC, Queensroad, Teddington, Middlesex TWY11 OLY, UK, ISBN 0-85404-432-9 (1997).

81. ISO/IEC Guide 43-1984: *Development and operation of laboratory proficiency testing* (1984). International Organization for Standardization, Case postale 56, CH-1211 Geneve 20, Switzerland.

82. Draft Revision of ISO/IEC Guide 43. Proposed title: *Proficiency testing by interlaboratory comparisons,* Committee Papers ILAC94, Hong-Kong, 17–21 October 1994, pp. 321–360.

83. M. Thompson and R. Wood, International harmonized protocol for proficiency testing of chemical analytical laboratories, *Pure Appli. Chem.* 65:2123–2144 (1993); published simultaneously in *J. AOAC Int.* 76 (4):926–940 (1993).

84. A. L. Patey, The Food Analysis Performance Assessment Scheme (FAPAS), *The VAM Bulletin* 11:12–13 (1994).

85. A. J. Trill, Computerized systems and GMP—a UK perspective: Part II, inspection findings, *Pharm. Tech.,* March 1993, pp. 49–62.

86. A. J. Trill, A regulatory perspective, in: *Computer Validation Practices,* Buffalo Grove, Ill., USA, Interpharm Press, Inc., ISBN 0-935184-5-4 (1994).

87. J. Guerra, Audits of computer systems in analytical laboratories, *Pharm. Tech.* 15 (9):142–148 (1988).

88. R. T. Tetzlaff, Validation issues for new drug development: Part II: Systematic assessment strategies, *Pharm. Tech.* Oct. 1992, pp. 84–94.

89. S. P. Bruederle, CGMPs and the pharmaceutical laboratory, presented at the HP seminar *Road to Compliance,* 1 March 1994, in Chicago.

90. R. Brown, FDA perspective on computer validation issues, presented at the Warner-Lambert Computer Validation Symposium, 30 April 1997, in Morris Plains, N.J., USA.

91. FDA 483 warning and violation letters, supplied by the CDER Freedom of Information Office, Rockville, ongoing sorted by month, http://www.fda.gov/cder/warn/index.htm.

92. EN 45020:1993, General terms and their definitions concerning standardization and related activities, Rue Brederode 2, B-1000 Brussels CEN/CENELEC, The Joint European Standard Institution.

93. M. Dorfman and R. Thayer, *Standards, guidelines, and examples on system and software requirements engineering*, Washington, D.C., IEEE Computer Society Press, ISBN 0-8186-8922-6 (1990).

94. J. S. Alford and F. L. Cline, PMA's Computer System Validation Committee, Computer System Validation—Staying current: Installation qualification, *Pharm. Tech.*, Sept. 1990, pp. 88–104.

95. D. L. Deitz and C. J. Herald, Reconciling a software development methothology with the PMA validation life cycle, *Pharm. Tech.*, June 1992, pp. 76–84.

96. Guy Wingate, *Validating automated manufacturing and laboratory applications: Putting principles into practice*, Interpharm Press, Inc., Buffalo Grove, Ill., USA, ISBN: 0- 57491-037-X (1997).

97. A. Brutsche, Calibration of analytical equipment (in German), presented at the seminar: Calibration and Qualification of Analytical Instruments in Pharmaceutical Quality Control, Concept Heidelberg, Darmstadt (Feb. 1998).

98. American Society for Testing and Materials. *The ASTM standard guide for training users of computerized systems* (1992).

99. K. Ishikawa, *Guide to quality control*, Nordica International Limited, ISBN 92-833-1035-7 (1990).

100. International Organization for Standardization (ISO), *Guide to the expression of uncertainty in measurement.* Case postale 56, CH-1211 Geneve 20, Switzerland 1993, ISBN 92-67-10188-9.

101. Operational qualification/performance verification for complete HP 1100 Series HPLC Systems, Hewlett-Packard Waldbronn, part number G2170-90102 (1996).

Index

A2LA. *See* American Association for Laboratory Accreditation
absorbance units, definition of, 211
acceptance criteria, definition of, 211
acceptance limits
    definition of, 63–64, 229
    graph of, 87
    for precision, 127
    setting, 72
accreditation, 221
    Accreditation for Chemical Laboratories: Guidance on the Interpretation of the EN 45000 Series of Standards and ISO/EC Guide 25, 30, 108
    definition of, 211
    National Voluntary Laboratory Accreditation Program (NVLAP), 222
accuracy
    of analytical methods, 129–130
    definition of, 211
AFNOR. *See* Association Francaise de Normalisation
AIA. *See* Analytical Instrument Association
American Association for Laboratory Accreditation (A2LA), 157, 211
American National Standards Institute (ANSI), 211
American Society for Quality (ASQ), 212
American Society for Testing and Materials (ASTM), 182, 212
analog/digital converter, operational qualification of, 243
analytical balances, operational qualification of, 238
Analytical Instrument Association (AIA), 22–23, 35
analytical methods, 127
    Analytical Methods Validation: Bioavailability, Bioequivalence and Pharmacokinetic Studies, 109
    limit of detection, 132–134, 135, 136
    limit of quantitation, 134–135
    linearity and calibration, 130–132, 135
    nonroutine methods, validation of, 119–120
    precision and reproducibility of results, 126–129
    robustness tests, 137
    stability of samples, 137–138
    standard methods, validation of, 115–119
    Standard Operating Procedure for testing of, 261–270
    validation of, 107–140
    validation parameters for, 123–140
    validation procedure, 109–115
Analytical Quality Control (AQC), 212
ANSI. *See* American National Standards Institute
AOAC. *See* Association of Official Analytical Chemists
The Application of the Principles of GLP to Computerized Systems, Series on Principles of Good Laboratory Practice and Compliance Monitoring, 30
application software, definition of, 212
AQC. *See* Analytical Quality Control
Asociación Latinoamericana de Integración, 211
ASQ. *See* American Society for Quality
assay, definition of, 212
Association Francaise de Normalisation (AFNOR), 211
Association of Official Analytical Chemists (AOAC), 211–212. *See also* Peer-Verified Methods Program
ASTM. *See* American Society for Testing and Materials
ASTM E 625-87: *Standard Guide for Training Users of Computerized Systems*, 182
audit, 5, 197–209. *See also* inspection guides
    checklist for, 205, 206–209
    definition of, 212
    horizontal, 201
    internal, 197, 220
    observations in, 198–200
    planning and implementation, 200–203
    report of, 203–204
    second-party, 197
    third-party, 197
    tracking, definition of, 212
    vertical, 201–202
audit tracking, definition of, 212

BAM. *See* Bundesanstalt fuer Materialpruefung und -forschung
BCR. *See* Community Bureau of Reference, Commission of the European Community
beta test, definition of, 212
biological fluids, analysis of, 109
black box testing, definition of, 212
BNF. *See* British National Formulary
BP. *See* British Pharmacopeia
British National Formulary (BNF), 212
British Pharmacopeia (BP), 213
British Standards Institute (BSI), 19, 213
Bundesanstalt fuer Materialpruefung und -forschung (BAM) (Germany), 166

caffeine, HPLC linearity plot of, 133
calibration, 13, 79
    curve for analytical methods, 130–132
    definition of, 32, 213
    of measuring equipment, 32
    vs. validation, 32
CANDA. *See* Computer-Assisted New Drug Application
capillary electrophoresis, definition of, 213
case studies of operational qualification, 283–286
CEN. *See* Comité Européen de Normalisation
Center of Drug Evaluation and Research (CDER), 2, 127
certification, 16, 214
certified reference material (CRM). *See also* reference material
    for accuracy assessment of equipment, 130
    classification of, 154
    definition of, 154, 213
    preparation and testing, 161–162
    quality control, use for testing, 84
    requirements for, 160–161
    use of, 164
CFR. *See* Code of Federal Regulations
change control, definition of, 213
checksum, definition of, 214
chemical industry, 109
chemistry, 10–11, 13
CITAC. *See* Co-operation on International Traceability in Analytical Chemistry
CITI. *See* Japanese Chemicals Inspection and Testing Institute
Co-operation on International Traceability in Analytical Chemistry (CITAC), 12, 214
Code d'Indexation des Materiaux de Reference (COMAR) (France), 167, 214
Code of Federal Regulations (CFR), 13, 214
COMAR. *See* Code d'Indexation des Materiaux de Reference
Comité Européen de Normalisation (CEN), 213

Community Bureau of Reference, Commission of the European Community (BCR), 212
compliance, definition of, 214–215
Computer-Assisted New Drug Application (CANDA), 213
computer database for reference materials (COMAR), 167
computer-related system, definition of, 215
computer software
    application software, definition of, 212
    assessment of, 18–19
    identification of, 55
    operational qualification of, 93–106
    source code, 225
    user-contributed software and macros, validation of, 103–104
    validation and qualification of, 3
    vendor-supplied, 95
Computer System Validation Committee of the PMA, 2, 33, 215
computer systems
    computer hardware, self-test during startup, 55, 56
    computer networks, operational qualification of, 99
    computerized analytical systems
        installation qualification, 56–57
        operational qualification of, 95–99
        Standard Operating Procedure for validation of, 275–281
    computerized system, definition of, 215
    definition of, 215
    identification of, 54–55
    installation qualification of, 53–57
    Laboratory Information Management Systems (LIMS), 95, 221
    operational qualification of, 93–106
    password, 222
    training end users, 182–185
    validation and qualification of, 4, 20
control charts, 83–89
Council Committee of Reference Materials of the ISO (REMCO), 225
cross-redundancy check of computer systems, 53
Current Good Manufacturing Practice for Drugs, General, 14

data
    reporting of, 144–145
    validation of, 141–144
debugging, definition of, 215
declaration of conformity, definition of, 215
Declaration of Operational Qualification, 249
Declaration of System Validation, 215
defective equipment, 89–91
Department of Health and Social Security (DHSS) (UK), 216

Index

design, definition of, 215
design qualification (DQ), 33–34, 41–47
design review, definition of, 216
design specifications, definition of, 216
Deutsches Arzneimittelbuch, 215
Deutsches Institut für Normung (DIN), 216
DHSS. *See* Department of Health and Social Security (UK)
DIN. *See* Deutsches Institut für Normung; Drug Information Number
diode-array detection in HPLC, 125
disaster recovery plan, definition of, 216
dissolution testing, operational qualification of, 244–245
documentation, 3–4
   analytical methods, validation report for, 114–115
   audit reports, 203–204
   calibration report, 79
   for computer systems, 51
   data reporting, 144–145
   Declaration of Operational Qualification, 249
   Declaration of System Validation, 215
   inspection guides, 11
   for installation qualification, 57–58
   job descriptions, 178
   logbooks, 75–77
   for operational qualification, 69, 104–105
   for performance qualification, 88–89
   personnel training records, 185–187
   of procedures, 3, 13–14, 21
   qualification of existing systems, 101
   of results, 3, 15
   Standard Operating Procedure, 225, 251–282
   test reports, 71
DQ. *See* design qualification
Drug Information Number (DIN), definition of, 216

EAC. *See* European Accreditation of Certification
EC. *See* European Community
EC Guide to Good Manufacturing Practice, 16, 219
EEC, 217. *See* European Community,
EFTA. *See* European Free Trade Agreement
e-mail address, author's, x, xi
EN 45000 series, 197
EN 45001 *General Criteria for the Operation of Testing Laboratories*, 10, 12, 18, 144–145, 217
Environmental Protection Agency (EPA), 156–157, 217
   Good Automated Laboratory Practice (GALP), 12, 219
   Guidance for Methods Development and Methods Validation for the Resource Conservation and Recovery Act (RCRA) Program, 2, 108
environmental testing
   operational qualification of HPLC for, 285–286
   pesticides in drinking water, HPLC of, 44–46

EOTC. *See* European Organization of Testing and Certification
EPA. *See* Environmental Protection Agency
equipment. *See also* computer systems; HPLC
   analog/digital converter, operational qualification of, 243
   analytical balances, operational qualification of, 238
   calibration of, 32
   characterization of, 53
   defective, handling of, 89–91
   definition of, 217
   design, 13
   design qualification, 41–43, 47
   dissolution testing, operational qualification of, 244–245
   flame atomic absorption spectrophotometer, operational qualification of, 238
   freezers, operational qualification of, 240
   furnaces, operational qualification of, 239
   gas chromatography, operational qualification of, 229–231
   infrared spectrometers, operational qualification of, 237
   installation qualification of, 50–52
   Karl Fischer apparatus, operational qualification of, 242
   maintenance, 13
   melting point apparatus, operational qualification of, 245
   modular vs. holistic testing, 66
   modules
      installation, 53
      replacement and operational qualification, 68
      self-test programs of, 53, 56
   operational qualification of, 61–72, 229–249
   performance qualification of, 73–91
   preinstallation qualification, 50
   qualification flowchart, 35
   qualification, timeline for, 34
   requalification, 68
   requirements for, 36
   Standard Operating Procedure for testing, 257
   sterilizers, hot air, operational qualification of, 240
   test equipment, 13
   tests during installation, 56–57
   validation and qualification of, 20, 21, 33–39
   viscosimeter, operational qualification of, 245
equipment modules, self-test programs of, 56
ERM. *See* external reference material
error, definition of, 217
ERS. *See* external reference specifications
escrow, definition of, 217
EU, 217. *See* European Community
EURACHEM, 2, 12, 20, 134, 136, 147

European Accreditation of Certification (EAC), 216
European Co-operation for the Accreditation of Laboratories (EAL), 20, 216
European Community (EC), 15, 17, 216
  EC Guide to Good Manufacturing Practice, 16, 219
  Good Manufacturing Practice, 29
  Good Manufacturing Practice for Medicinal Products in the European Community, 14–15
  The Harmonization of Laws, Regulations and Administrative Provisions to the Application of Good Laboratory Practice and the Verification of Their Application for tests on Chemical Substances, 13
  The Inspection and Verification of Good Laboratory Practice, 13
  Standards, Measurement and Testing Programme, 157, 165
European Economic Community (EEC), 217
European Free Trade Agreement (EFTA), 216
European Organization of Testing and Certification (EOTC), 217
European Pharmacopeia, 217
European Standards Institution, 213
European Union (EU), definition of, 217–218
existing systems, validation and qualification of, 100–102
external audit, definition of, 218
external reference material (ERM), 154
external reference specifications (ERS), definition of, 218

FDA. See Food and Drug Administration
Federal Food, Drug, and Cosmetic Act, 219, 226
Federal Insecticide, Fungicide and Rodenticide Control Act (FIFRA), 13
FIFRA. See Federal Insecticide, Fungicide and Rodenticide Control Act
firmware, definition of, 218–219
flame atomic absorption spectrophotometer, operational qualification of, 238
flowcharts. See also procedures
  analytical laboratories, validation and qualification of, 5
  analytical procedures, validation of, 3, 118
  data checking, 143
  equipment, handling defective, 90
  equipment qualification, 4, 34, 35
  laboratory proficiency testing, 190
  optimization of validation, 7
  quality control (QC) samples with control charts, 84
  quality pyramid, 24
  revalidation, 123
  stability testing, 138
  standard methods, evaluation and validation of, 117

Standard Operating Procedure for equipment testing, 257
testing, calibration, qualification, verification, and validation, 31
uncertainty, estimation of, 148
validation and qualification, principles of, 28
Food and Drug Administration (FDA), 218, 219, 224
  audits, laboratory, 198
  Compliance Policy Guide, 218
  Current Good Manufacturing Practice for Drugs, General, 14
  Current Good Manufacturing Practice for Finished Pharmaceuticals, 14
  Good Laboratory Practice, 12
  Good Manufacturing Practice, 108
  Guide to Inspection of Pharmaceutical Quality Control Laboratories, 16
  guidelines for analytical methods, 108
  inspection guides, 11
  Inspectors Technical Guide, 218
  modular vs. holistic operational qualification, 66
  recommendations, definition of, 218
  Standard Operating Procedure, 225
  Technical Review Guide: Validation of Chromatographic Methods, 2
  U.S. Pharmacopeia, 226
  vendor qualification and, 43
  warning letter, 226
  web page, 11
freezers, operational qualification of, 240
frequency of operational qualification testing, 65–66
functional specification, definition of, 219
furnaces, operational qualification of, 239

GALP. See Good Automated Laboratory Practice
GAMP. See Good Automated Manufacturing Practice
GAP. See Good Analytical Practice
gas chromatography, operational qualification of, 229–231
GCP. See Good Clinical Practice
General Criteria for the Operation of Testing Laboratories, 18
General Principles of Validation, 28
General Requirements for the Competence of Calibration and Testing Laboratories, 18
GLP. See Good Laboratory Practice
GMP. See Good Manufacturing Practice
Good Analytical Practice (GAP), 6, 219
Good Automated Laboratory Practice (GALP), 12, 219
Good Automated Manufacturing Practice (GAMP), definition of, 219
Good Clinical Practice (GCP), 219
Good Laboratory Practice (GLP), 1, 10, 11, 12–14, 23–24
Good Laboratory Practice in the Testing of Chemicals, 13, 14
Good Manufacturing Practice (GMP), 1, 10, 11, 14–16, 214, 219–220

Index

Current Good Manufacturing Practice for Drugs, General, 14
EC Guide to Good Manufacturing Practice, 16
Good Manufacturing Practice for Medicinal Products in the European Community, 14–15
Good Pharmaceutical Manufacturing Practices (UK), 225–226
Guide to Good Manufacturing Practice for Pharmaceutical Products, 15
International Biotechnology, Bulk Chemical, and Pharmaceutical GMPs, 15
pharmaceutical industry, 23–24, 225
Guidance for Methods Development and Methods Validation for the Resource Conservation and Recovery Act (RCRA) Program, 2
Guide to Inspection of Pharmaceutical Quality Control Laboratories, 16
guidelines, 9–25
Food and Drug Administration, definition of, 218
USP vs. ICH for validation of analytical methods, 108, 139, 140
validation experiment sequence, 114
Guidelines for the Application of ISO 9001 to the Development, Supply and Maintenance of Software, 19

Handbook for Inspectors in Germany, 11
The Harmonization of Laws, Regulations and Administrative Provisions to the Application of Good Laboratory Practice and the Verification of Their Application for Tests on Chemical Substances, 13
Hewlett-Packard HP1050 Series HPLC, Standard Operating Procedure for validating precision of, 270–275
Hewlett-Packard HP1100 Series HPLC, 66–67, 71, 215, 285–286
high performance liquid chromatography. See HPLC
HPLC, 88–89
of caffeine, 133
diode-array detection in, 125
for environmental testing, 285–286
Hewlett-Packard HP1100 Series HPLC, 66–67, 71, 215, 270–275, 285–286
logbook for, 76
operational qualification of, 96–97, 234–236, 283–284
pure and impure peaks, 126
qualification steps for, 44–46, 65
selectivity and specificity, validation of, 125–126
Standard Operating Procedure for validating precision of, 270–275
human resources. See personnel
IAEA. See International Atomic Energy Agency
ICH. See International Conference on Harmonization . . .

ILAC. See International Laboratory Accreditation Conference (ILAC)
implementation
of analytical methods validation, 121
of computer systems and software, 104–105
of operational qualification, procedure for, 70–71
Information Quality Technology System (IQTS), 18–19
infrared spectrometers, operational qualification of, 237
inspection, definition of, 220
The Inspection and Verification of Good Laboratory Practice, 13
inspection guides, 11, 16
installation qualification (IQ), 33–34, 49–59, 220–221
module qualification, 53
protocol for, 57–58
requalification after system changes, 59
verification of, 57, 58
instrumentation. See equipment
interlaboratory test comparisons, definition of, 220
intermediate precision, 127–128
internal audit, definition of, 197, 220
internal reference material (IRM), 154, 163–164
International Atomic Energy Agency (IAEA), 167
international companies and multiple regulations, 23–25
International Conference on Harmonization of Technical Requirements for Registration of Pharmaceuticals for Human Use (ICH), 12, 16–17, 108, 212, 220
analytical method validation characteristics, 139, 140
intermediate precision, definition of, 127
linear curves and accuracy reporting, 132
precision and reproducibility of results, 126
range, minimum specified range, 132
reproducibility, definition of, 128
specificity/selectivity, definition of, 124–125
Validation of Analytical Procedures: Definitions and Terminology (Q2A), 17
Validation of Analytical Procedures: Methodology (Q2B), 17
International Guide to Quality in Analytical Chemistry: An Aid to Accreditation, 19–22, 30, 79–80, 81, 116, 120
International Laboratory Accreditation Conference (ILAC), 20, 220
International Organization for Standardization (ISO), 2, 221
Council Committee of Reference Materials of the ISO (REMCO), 225
Guide to the Expression of Uncertainty in Measurement, 147
Guidelines for the Application of ISO 9001 to the Development, Supply and Maintenance of Software, 19

ISO 7870 *Control Charts: General Guide and Introduction*, 83, 85
ISO 8258 *Shewhart Control Charts*, 85–87
ISO 9000, 10, 18–19, 197, 221
ISO 9000-3, 18–19, 33
ISO 9001, 19, 23–25
ISO 9003, 186
ISO Council Committee on Reference Materials (REMCO), 166, 225
ISO/DIS 254-1, 226
ISO/IEC Guide 25, 10, 18, 23–24, 30, 31, 32, 220, 221
  data reporting, 144–145
  measurement of uncertainty, 146–147
  reference standards, 156
ISO/IEC Guide 33: Uses of Certified Reference Materials, 154
ISO/IEC Guide 35: Certification of Reference Materials: General and Statistical Principles, 151
ISO/IEC Guide 43, 190
international standard, definition of, 220
International Union of Pure and Applied Chemistry (IUPAC), 20, 124, 167
Internet
  e-mail address, author's, x, xi
  newsgroups, 221
  uniform resource locator, 226
  web pages, x, xi, 11, 222
  world wide web, 226
interviews, personnel, procedure for, 173–178
IQ. *See* installation qualification
IQTS. *See* Information Quality Technology System
IRM. *See* internal reference material
ISO. *See* International Organization for Standardization
IUPAC. *See* International Union of Pure and Applied Chemistry

Japan, 11, 17, 158
Japanese Chemicals Inspection and Testing Institute (CITI), 158
Japanese Pharmacopeia (JP), 221

Karl Fischer apparatus, operational qualification of, 242

Laboratory Information Management Systems (LIMS), 95, 221
Laboratory of the Government Chemist (LGC), 166. *See also* LGC/EURACHEM
Laborattoire National d'Essais (LNE) (France), 166
legislation, 9
  European Community and, 13
  Federal Food, Drug, and Cosmetic Act, 219
  Federal Insecticide, Fungicide and Rodenticide Control Act (FIFRA), 13

Resource Conservation and Recovery Act (RCRA), 2, 108
Toxic Substances Control Act (TSCA), 13
LGC. *See* Laboratory of the Government Chemist; LGC/EURACHEM
LGC/EURACHEM, 2, 22, 63
  maintenance and operational qualification, 67
  modular vs. holistic operational qualification, 66
  qualification guidelines, 34–35
limit of detection (LOD)
  in analytical methods, 132–134, 135, 136
  definition of, 221
limit of quantification (LOQ), definition of, 134–135, 221
LIMS. *See* Laboratory Information Management Systems
linearity of analytical methods, 130–132, 135
LNE. *See* Laborattoire National d'Essais
LOD. *See* limit of detection
logbooks, 75–77
LOQ. *See* limit of quantification

maintenance
  of equipment, 13
  and operational qualification, 67–68
  and performance qualification, 77–78
  repairs and operational requalification, 67–68, 90
manufacturing. *See* production
melting point apparatus, operational qualification of, 245
Ministry of Health, definition of, 221
modules. *See* computer systems

NACCB. *See* National Accreditation Council for Certification Bodies
NAMAS. *See* National Measurement Accreditation Service
National Accreditation Council for Certification Bodies (NACCB), 226
National Bureau of Standards (NBS), 221
National Institute for Environmental Studies (NIES) (Japan), 166
National Institute of Materials and Chemical Research (NIMC) (Japan), 158
National Institute of Standards and Technology (NIST), 20, 154, 156, 166, 222
National Institute of Technology and Evaluation (NITE) (Japan), 158
National Measurement Accreditation Service (NAMAS), 19, 217–218, 221, 226
National Research Center for Certified Reference Materials (China), 166
National Research Council (NRC) (Canada), 166
national standard, definition of, 221
National Technical Information Service (NTIS), 222

National Voluntary Laboratory Accreditation Program (NVLAP), 222
NBS. *See* National Bureau of Standards
newsgroups, definition of, 222
NIES. *See* National Institute for Environmental Studies
NIMC. *See* National Institute of Materials and Chemical Research (Japan)
NIST. *See* National Institute of Standards and Technology
NITE. *See* National Institute of Technology and Evaluation
NRC. *See* National Research Council (Canada)
NVLAP. *See* National Voluntary Laboratory Accreditation Program

OECD. *See* Organization for Economic Cooperation and Development
Office of Reference Material (ORM) (UK), 166
operational qualification (OQ), 33–34, 222
    acceptance limits, 229
    acceptance limits and, 63–64
    of analog/digital converter, 243
    of analytical balances, 238
    case studies on, 283–286
    of computer networks, 99
    of computer systems, 93–106
    of computer systems and software, 97
    Declaration of Operational Qualification, 249
    of dissolution testing, 244–245
    documentation and archives, 229–230
    of equipment, 61–72
    of flame atomic absorption spectrophotometer, 238
    of freezers, 240
    frequency of, 65–66, 70
    of furnaces, 239
    of HPLC, 234–236, 283–286
    implementation, procedure for, 70–71
    of infrared spectrometers, 237
    of Karl Fischer apparatus, 242
    of melting point apparatus, 245
    modular vs. holistic testing, 66
    moving equipment and requalification, 68
    of ovens, 239
    of pH meter, 246
    of polarimeter, 247–248
    procedure for, 62–63
    of refractometer, 247
    of refrigerators, 240
    requalification after system changes, 67–68, 90
    of software, 93–106
    of spectrophotometers, 232–233
    tests for selected equipment, 229
    of thermometers and thermocouples, 241
    traceability, 229
    of viscosimeter, 245
optimization of validation, 6–7
OQ. *See* Operational qualification
Organization for Economic Cooperation and Development (OECD), 169, 222, 226
    Good Laboratory Practice in the Testing of Chemicals, 13, 14
    operational qualification of computer systems and software, 93
ORM. *See* Office of Reference Material
out-of-control events, 87, 89, 222
ovens, operational qualification of, 239

PASG. *See* Pharmaceutical Analysis Science Group
password, definition of, 222
Peer-Verified Methods Program, 2, 109, 127, 130
performance qualification (PQ), 33, 35, 73–91
    definition of, 73, 222–223
    frequency of, 74, 80, 82
    out-of-control events, 87, 89
    procedure for, 81
    system suitability testing, 80
performance verification (PV), definition of, 223
personnel, 169–188
    ASTM E 625-87: *Standard Guide for Training Users of Computerized Systems*, 182
    job descriptions, 178–179
    recruiting and interviewing, 170–178
    training, 179–187
pesticides in drinking water, HPLC of, 44–46
pH meter, operational qualification of, 246
Pharmaceutical Analysis Science Group (PASG) (UK), 2, 23, 222
pharmaceutical industry, 11, 109
    Computer System Validation Committee, 2, 33, 215
    marketing, international, 23–25
    operational qualification of HPLC for, 283–286
Pharmaceutical Industry Computer Systems Validation Forum (PICSVF), 223
Pharmaceutical Inspection Convention (PIC), 15, 223
Pharmaceutical Manufacturers' Association (PMA), 33, 63, 223
pharmaceuticals
    Computer-Assisted New Drug Application (CANDA), 213
    Current Good Manufacturing Practice for Drugs, General, 14
    Good Pharmaceutical Manufacturing Practices (UK), 225–226
    Guide to Good Manufacturing Practice for Pharmaceutical Products, 15
    International Biotechnology, Bulk Chemical, and Pharmaceutical GMPs, 15

International Conference on Harmonization of Technical Requirements for Registration of Pharmaceuticals for Human Use (ICH), 16–17
   operational qualification of HPLC for, 283–284
Pharmaceutical Analysis Science Group (PASG) (UK), 2, 23, 222
Pharmaceutical Industry Computer Systems Validation Forum (PICSVF), 223
Pharmaceutical Inspection Convention (PIC), 15, 223
Pharmaceutical Manufacturers Association (PMA), 223
pharmacopeia, definition of, 223
   Standard Operating Procedure, 225–226
   U.S. Pharmacopeia, 17, 226–227
   validation of, 30
pharmacopeia. See also U.S. Pharmacopeia (USP)
   British Pharmacopeia, 213
   definition of, 223
   Deutsches Arzneimittelbuch, 215
   European Pharmacopeia, 217
   Japanese Pharmacopeia, 221
PIC. See Pharmaceutical Inspection Convention
PICSVF. See Pharmaceutical Industry Computer Systems Validation Forum
PMA. See Pharmaceutical Manufacturers Association
polarimeter, operational qualification of, 247–248
PQ. See performance qualification
precision, 126–129
   definition of, 223
   intermediate, 127–128
   Standard Operating Procedure for validation of precision of HPLC, 270–275
preinstallation qualification, procedure for, 50
preventive maintenance. See maintenance
primary standard, 154
procedures. See also flowcharts; Standard Operating Procedure (SOP)
   audits, laboratory, 200–203
   certified reference materials, preparation and testing of, 161–162
   data validation, 142–144
   for design qualification, 42
   equipment characterization, 53
   evaluation of uncertainty, 147–149
   existing systems, validation and qualification of, 100–102
   implementation of operational qualification, 70–71
   for installation qualification, 52
   internal reference materials, preparation of, 163–164
   operational qualification, 62–63
      of analog/digital converter, 243
      of analytical balances, 238
      for capillary electrophoresis, 231
      of computer hardware and software, 95–99
      of dissolution testing, 244–245
      of flame atomic absorption spectrophotometer, 238
      of freezers, 240
      of furnaces, 239
      for gas chromatography, 229–231
      of HPLC, 234–236, 284–286
      of infrared spectrometers, 237
      of Karl Fischer apparatus, 242
      of melting point apparatus, 245
      of ovens, 239
      of pH meter, 246
      of polarimeter, 247–248
      of refractometer, 247
      of refrigerators, 240
      of spectrophotometers, 232–233
      of thermometers and thermocouples, 241
      of viscosimeter, 245
   for performance qualification, 81
   personnel, recruiting and interviewing, 170–178
   for preinstallation qualification, 50
   proficiency testing, laboratory, 190–191
   for qualification, 34, 35
   Standard Operating Procedure, 251–282
   for validation and qualification, 37–39
   validation of methods, 110
   validation of standard methods, 116–117
   validation of user-contributed software, 103–104
   for vendor qualification, 43, 47, 48
   working standards, preparation of, 163
production, 16
proficiency testing
   definition of, 223
   frequency of, 192–193
Proficiency Testing in Analytical Chemistry, 189
prospective validation, definition of, 223
prototyping, definition of, 224

QA. See quality assurance
QC. See quality control
qualification, 3–5. See also validation
   definition of, 33–35, 224
   design qualification, 33–34, 41–47
   development and implementation, 35, 37–39
   flow chart of, 5
   installation qualification, 33–34, 49–59, 220–221
   operational qualification, 33–34, 61–72, 222
   performance qualification, 33, 35, 73–91, 222–223
   preinstallation, 50
   procedure for, 37–39
   timeline for, 34
   of vendors, 43, 47, 48
   vs. validation, 33
quality assurance program, laboratory, 165

quality assurance (QA), definition of, 224
quality control (QC), 22. *See also* qualification; validation
    definition of, 224
    Good Manufacturing Practice and, 15–16
    ISO 8258 *Shewhart Control Charts*, 85–86
    plan for, 120–121
    proficiency testing, laboratory, 189–195
    samples with control charts, 83–89
quality pyramid, 24
quality standards, international, meeting, 23–25
Quantifying Uncertainty in Analytical Measurement, 147

R-charts, 85
range, of analytical methods, 132
raw data, definition of, 224
RCRA. *See* Resource Conservation and Recovery Act
recovery, of analytical methods, 130
reference material, 153–154. *See also* standards
    availability of, 165–167
    computer database (COMAR), 166
    definition of, 224
Reference Materials Advisory Service (REMAS) (UK), 167
reference standard, definition of, 224
refractometer, operational qualification of, 247
refrigerators, operational qualification of, 240
regulations, 9–25
regulatory methods validation, definition of, 225
relative standard deviation, 83, 148
REMCO. *See* Council Committee of Reference Materials of the ISO
repairs. *See* maintenance
reproducibility, 128–129
requalification. *See* operational qualification (OQ)
Resource Conservation and Recovery Act (RCRA), 2, 108
retrospective validation, definition of, 225
revalidation, definition of, 225. *See also* validation
robustness tests of analytical methods, 137
ruggedness tests, 135, 137, 225
Rules Governing Medicinal Products in the European Community, 219

sampling, 3
    quality control samples with control charts, 83–89
    stability of samples, 137–138
second-party audits, 197
secondary standard, 154
selectivity, of analytical methods, 124–126
Shewhart Control Charts, 85–86
SI. *See* System International
software. *See* computer software
SOP. *See* Standard Operating Procedure
source code, definition of, 225
specificity, of analytical methods, 124–126

spectrophotometers, operational qualification of, 232–233
SRM. *See* standard reference material
staff. *See* personnel
standard deviation of limits of detection, 134
Standard Operating Procedure (SOP), 13–14. *See also* procedures
    definition of, 225–226
    for equipment and procedures, 251–282
    Standard Operating Procedure, writing, 258–260
    for validating computerized analytical systems, 275–282
    for validating precision of HPLC, 270–275
    for validation of analytical methods, 261–269
standard reference material (SRM), 154
standards, 9–25, 65, 72
    certified reference material, 84, 129–130, 158–160
    internal reference material, 154, 163–164
    international standards, definition of, 220
    laboratory reference standards, 161
    national, definition of, 221
    official reference material programs, 156–158
    quality assurance program, laboratory, 165
    reference standards, 151–167
    traceability, 155, 158–160, 229
    working standard, 154, 163
sterilizers, hot air, operational qualification of, 240
System International (SI), definition of, 225
system stability, definition of, 137–138
system suitability testing, 80, 82–83, 226

Technical Review Guide: Validation of Chromatographic Methods, 2
terminology of validation, 27–30
test, definition of, 226
test plan, definition of, 226
testing
    alpha test, definition of, 211
    beta test, definition of, 212
    black box testing, definition of, 212
    computer hardware, self-test during startup, 53, 56
    definition of, 31
    flowchart for Standard Operating Procedures, 257
    proficiency testing, laboratory, 189–195
    vs. validation, 31–32
TGA. *See* Therapeutic Goods Administration
theophylline, HPLTC of, 109
Therapeutic Goods Administration (TGA) (Australia), 226
thermometers and thermocouples, operational qualification of, 241
third-party audits, 197
TickIT, 18–19
time vs. usage basis for maintenance and performance requalification, 77–78
timelines, equipment qualification, 34

toxicology, 12–13
traceability, 155, 158–160, 226, 229
training, personnel, 179–187
    ASTM E 625-87: *Standard Guide for Training Users of Computerized Systems*, 182
    by vendors, 183
TSCA. *See* Toxic Substances Control Act

UKAS. *See* United Kingdom Accreditation Service
UK Pharmaceutical Analysis Science Group. *See* Pharmaceutical Analysis Science Group
uncertainty, 146–149, 226
uniform resource locator (URL), 226
United Kingdom, 11, 19
United Kingdom Accreditation Service (UKAS), 9, 226
United States, 11, 17, 156–157
upgrades of systems, and operational qualification, 67–68
URL. *See* uniform resource locator
U.S. Pharmacopeia (USP), 12, 17, 226–227
    analytical method validation characteristics, 140
    ruggedness, definition of, 135, 137
    selectivity, definition of, 125
    system suitability testing, 82–83
    validation, definition of, 30
    validation procedure, 108
USP. *See* U.S. Pharmacopeia

validation, ix–xi, 1–7, 27–39. *See also* qualification
    of analytical methods, 107–140
    audits, laboratory, 197–209
    of computer systems and software, 93–105
    of data, 141–144, 215
    Declaration of System Validation, 215
    definition of, 28–30, 227
    evaluation of uncertainty, 147–149
    flow chart of, 5
    General Principles of Validation, 28
    The Inspection and Verification of Good Laboratory Practice, 13
    of nonroutine analytical methods, 119–120
    optimization of, 6–7
    parameters for analytical method validation, 112, 123–140
    procedure for, 37–39
    prospective, definition of, 223
    regulatory methods, 224
    retrospective, 224
    revalidation of analytical methods, 122–123
    of software, user-contributed, 103–104
    of standard analytical methods, 115–119
    Standard Operating Procedures for, 251–282
    steps of, 4
    Technical Review Guide: Validation of Chromatographic Methods, 2
    terminology of, 27–30
    Validation Master Plan, 111
    Validation of Analytical Procedures: Definitions and Terminology (Q2A), 17
    Validation of Analytical Procedures: Methodology (Q2B), 17
    Validation of Computerized Analytical Systems, x, xi, 94
    validation protocol, definition of, 227
    vs. verification, testing, calibration, and qualification, 31–35
validation protocol, definition of, 227
vendors
    operational qualification of equipment by, 66–67
    personnel training by, 183
    qualification of, 43, 47, 48
verification
    of data from computerized analytical systems, 97
    definition of, 32–33, 226
    performance, 222
    report of, 71
    vs. validation, 32–33
viscosimeter, operational qualification of, 245
visual inspection of detection limits, 134

warning letter (FDA), definition of, 227
water, HPLC of pesticides in, 44–46
WELAC. *See* Western European Laboratory Accreditation Conference
Western European Laboratory Accreditation Conference (WELAC), 12, 227
WHO. *See* World Health Organization
working standard, 154, 163
World Health Organization (WHO), 15, 29
world wide web, 226

X-charts, 85–86